T0179888

PRIMATE SOCIOECOLOGY

PRIMATE SOCIOECOLOGY

Shifting Perspectives

Lynne A. Isbell

Johns Hopkins University Press
Baltimore

Johns Hopkins University Press
2715 North Charles Street
Baltimore, Maryland 21218
www.press.jhu.edu

Library of Congress Cataloging-in-Publication Data

Names: Isbell, Lynne A., author.
Title: Primate socioecology : shifting perspectives / Lynne A. Isbell.
Description: Baltimore : Johns Hopkins University Press, [2024] |
    Includes bibliographical references and index.
Identifiers: LCCN 2023033226 | ISBN 9781421448909 (hardcover ;
    acid-free paper) | ISBN 9781421448916 (ebook)
Subjects: LCSH: Primates—Behavior. | Social behavior in animals.
    Classification: LCC QL737.P94 I83 2024 | DDC 599.88—dc23/eng
    /20231208
LC record available at https://lccn.loc.gov/2023033226

A catalog record for this book is available from the British Library.

Special discounts are available for bulk purchases of this book. For more
information, please contact Special Sales at specialsales@jh.edu.

CONTENTS

TABLES

PREFACE

Ah, primate socioecology, the love of my professional life. It is an important field of study for those who are interested in how natural selection operates on organisms as they interact with their environments. If we can understand those interactions, we will be able to grasp more clearly how the natural world works. Practically speaking, that knowledge should help us predict what will come next for animals as the world's climate changes, so that with intervention of some kind, we might help to minimize animal extinctions.

Socioecology grabbed hold of me more than 45 years ago, and it has not let me go. I have been thinking about it, trying to figure it out, since I moved to Davis, California, in 1977 and sat in on one of Peter Rodman's classes. Why do primates have the diversity of social organizations that they have? What causes some to live alone, others to live in pairs, and still others to live in large groups? Why are so many nocturnal primates solitary foragers? Why are so many nocturnal primates small-bodied? Why do most diurnal primates live in cohesive groups? Why are orangutans the only diurnal primate to forage solitarily? Admittedly, these are not the kinds of questions that are foremost in the minds of most people and not what a popular audience cares about, but they have persisted in my mind even when most other socioecologists have settled on answers that they find satisfying. This book offers a new perspective because in all the years that socioecology has been around as a field of inquiry, we socioecologists still have not been able to answer those basic questions to my satisfaction. To make progress, I believe we need to return to the starting point and try a new direction.

I thought I had finally succeeded in satisfying at least myself when, back in 2004, I proposed the Dispersal / Foraging Efficiency model (Isbell 2004). I came to it by way of reading theoretical and empirical studies and through my own fieldwork on three species of African primates, which focused on ranging behavior, food competition, predation, and dispersal. Following in the footsteps of socioecological models of the past, I placed females into different categories, with the categories based on a novel angle. Rather than seeing dispersal from the perspective of the offspring, I turned it to the mother's perspective and asked how willing mothers are to share their home ranges with their adult daughters and how much their home ranges overlap with other groups.

During the intervening years, I kept scrutinizing the model, and I became dissatisfied again—too many species seemed to be exceptions. With more thinking and reading and thinking again, I revised that model to become the Variable Home Range Sharing model, which I present in this book. I still come at it from the mother's perspective, but female categories have been revised based on a simple dichotomous key that allows us to classify females easily enough, as long as we know if females control their home range boundaries and, for those who do, if we know (1) the extent of home range sharing with other reproductive-aged females; (2) the mechanism of female dispersal (no dispersal, or dispersal that is either forced or volitional); and (3) whether other factors, such as inbreeding, infanticide, or predation, hint at having a greater effect than food on female reproductive success.

From the information currently available, I am confident that the Variable Home Range Sharing model explains variation in primate social organizations better than any other previous model. I believe that I am finally satisfied. Still, I do not hope for the same from you after you have read and understood the model because science cannot advance that way. What I do hope is that you will become convinced enough of the model's potential that you will subject it to further scientific investigation.

ACKNOWLEDGMENTS

There have been three phases to this book: (1) *Incubation*, which took 45 years; (2) *Formulation*, which took 2 more years; and, finally, (3) *Dissemination*.

For the Incubation phase, I want to give credit to the many people who got me started. Gerry Gates, professor of biology at the University of Redlands, loaned me his copy of E. O. Wilson's *Sociobiology* before he'd even had a chance to read it. I'm sorry it got wet when a friend's rusty old Valiant I was riding in went through the car wash. Its poor treatment was not a reflection on how I felt about the field. Joe Skorupa showed me by example how graduate school works and provided companionship in Uganda's Kibale Forest Reserve (now Kibale National Park) during my first multiyear stint with fieldwork, an opportunity that never would have happened without his willingness to switch temporarily from avian to primate ecology. He also helped me through the process of publishing my first paper. When it was published, I figured I would stop there since my name on a publication meant that I was now immortal in a way. I thought, *What more than immortality could anyone want?*

But then a group of University of California, Davis, faculty members, including Ben Hart and Dick Coss, decided to develop a new graduate program in animal behavior. This was highly fortuitous for me—all I'd ever wanted to do was study animal behavior, and now there was a graduate program in the town where I was living that might allow me to do that for longer. Ben actually paid me to observe domestic cat behavior in his lab before the UC Davis Animal Behavior Graduate Group began, and Dick advocated for my admission into it after I returned from Kibale.

Tom Struhsaker, the best teacher in the field any budding primatologist could have, taught me how to identify different tree species and individual red colobus monkeys. He also taught me to always whisper in the forest, rest my binoculars flat on a table, and respect the forest and the data.

Peter Rodman was critical to my primatological growth. Without his contacts and support, I would not have been able to volunteer for a year to collect behavioral data on bonnet macaques at the California Regional Primate Center (now called the California National Primate Research Center, or CNPRC). Nor would I have been able to spend two years studying red colobus in Kibale and another two years studying vervets in Amboseli National Park. He also offered to serve as my major professor in graduate school. With sociobiological theory still new, it was an exciting time to be in grad school, and with his precise thinking, Peter made sure his students became well-versed in the theory. I think he did more to advance primatology through his teaching than anyone could by writing a book.

Although I had enough data from the red colobus for a dissertation, the "Africa bug" had bitten me, and I desperately wanted to return. It was more important to me to experience more fieldwork and other primate species than to go through graduate school quickly. Robert Seyfarth and the late Dorothy Cheney gave me that opportunity when they hired me as field manager for their long-term project in Amboseli, an open savannah environment where I felt most at home. I ended up writing my dissertation on the Amboseli vervets. Amboseli is also where I met Truman Young, who helped me through the last stage of graduate school by critically reading drafts of my dissertation chapters and giving me tips on how to apply for faculty positions. Extending well beyond that stage of my career, I benefited immensely from his uncanny ability to cut through to the important issues in any discussion. Together, we set up our next field site in Laikipia, Kenya (heaven on Earth), and later enjoyed showing our son the many beautiful aspects of East Africa. What an adventure! Those memories now get to live beyond us into the next generation.

My research over the years helped the incubation of the ideas in this book. While the research projects were planned, I learned so much more from simply observing animals that were different in one way or another: red colobus, which have female dispersal and seemed to have plenty of food to eat no matter the size of their groups; patas monkeys, which seemed to always be on the move; and vervets, which never seemed to get a break from leopard predation. Habitat deterioration and intense predation, events that couldn't be planned at my study sites, led me to think more carefully about the costs of dispersal and to view dispersal from the mother's perspective.

The Incubation phase shifted to the Formulation phase when I was able to devote my sabbatical year from UC Davis to working on this book. I am very grateful to all the primatologists over the years who put in the hard work required to publish their findings, which made it possible for me to synthesize the material. Dorothy Fragaszy and Sally Mendoza patiently handled my numerous questions about the species of squirrel monkeys kept at the CNPRC in the 1970s and early 1980s, and I thank them for sharing their knowledge. Feedback from Peter Rodman on the history of socioecology and Cynthia Thompson on thermoregulation helped strengthen chapters 1 and 7, respectively. Comments on the entire manuscript from Krishna Balasubramaniam, Truman Young, and the graduate students in my fall 2022 seminar—Nalina Aiempichitkijkarn, Bradley Kristin, Emily Monroe, and Abigail Morris—greatly improved earlier drafts. Thanks very much for taking the time to make this book better.

Tiffany Gasbarrini, senior acquisitions editor at Johns Hopkins University Press, was such a champion of this book that at times my eyes welled up with tears of gratitude for her enthusiastic support. She never wavered in her optimism as she ushered the manuscript through the various stages of approval. One of those stages involved finding six people who would agree to evaluate the book manuscript's suitability for publication. I thank those anonymous reviewers for their time and positive comments. Ezra Rodriguez, assistant acquisitions editor, then deftly moved the book manuscript to the next level by helping me turn the various files into acceptable formats for the production and design folks.

Kathy West gets a big hug and glowing appreciation for giving me Sammy, the happiest dog I've ever known, and for illustrating the book. Our collaborations always help me to see my own ideas more clearly. I'm grateful that the UC Davis Academic Senate Faculty Committee on Research saw value in providing funds for her to create the illustrations.

The Johnston Manufacturers (you know who you are) get a call-out for keeping things light every day during the Formulation phase. My gratitude to you all for the laughs and unconditional support. Special thanks to fellow Manufacturer Freeman Tinnin for checking on me every night to make sure I'm still alive, and to fellow Manufacturer and old friend Lennie Denevan for strategic discussions about the book and because I forgot to thank her in my previous book. I'm also grateful to Harry Greene for his unwavering encouragement; to the Davis pickleballers for the happiness they gave me between bouts of writing, bringing home to me the importance of "weak ties"; and to Bonita Bradshaw, Hal Hinkle, Jan Kalina, Jeanetta Mastron, Becky Pantaleon, Sharleen Pearson, and Joe Skorupa for not giving up on me. This book is dedicated to my son, Peter Young, my greatest joy.

With Dissemination now at hand, I get to thank the design team at Johns Hopkins University Press for their efforts in creating an attractive book, production editor Hilary Jacqmin for walking me through the editing process, and copyeditor Paul Payson for his sensitivity and extraordinarily diligent attention to detail. So many commas, deleted here and added there! I thought I'd given him a grammatically correct and punctuationally perfect manuscript, but his ability to catch my mistakes have humbled me. I doubt there are any mistakes left but if there are, they are mine alone. Thanks to you all for helping this book see the light of day.

PRIMATE SOCIOECOLOGY

CHAPTER 1

Highlights in the History of Primate Socioecology

We need, in effect, to develop new ways of describing
primate societies that break the old moulds within which
our thinking has been constrained.

—R. I. M. Dunbar (1988:11)

Those who study animal behavior follow a basic approach. In essence, we see what animals do, and then we try to explain it. People like me, who focus on primate socioecology (a field of study integrating animal behavior and ecology), try to explain it starting with the premise that, in general, female primates are most limited by food, and male primates are most limited by access to females as mates. An emergent property arising from that premise is their "social organization." Primates exhibit a range of recognizable social organizations, including living as solitary foragers or in "fission-fusion" communities whose members routinely come together and go their separate ways over the course of days, weeks, or months. Other more socially cohesive options include living in pairs; in single-male, multi-female groups; or in multi-male, multi-female groups.

To explain this diversity, we need more than that basic premise, and so we have focused on primate responses to potential predators and to various aspects of their food, particularly the types of foods eaten, their abundance, and their spatial or temporal distribution, which I will argue has taken us down the wrong path. To understand how we went in that direction and how it differs from the new direction I am proposing we follow, it will help to know a little of the history of socioecological theory.

The Initial Framework (Crook and Gartlan 1966)

John Crook, an ornithologist by training, studied weaver birds, and he thought that the diversity in their nest shapes could be explained by environmental conditions, including predation and weather (Crook 1963). This line of thinking was then applied to explain the diversity in avian social organizations, with the main selective agents being habitat, type of foods eaten, food abundance and distribution, and predation—particularly in nest site selection (Crook 1965). A year later, Crook teamed up with primatologist Steve Gartlan to apply the same logic to primates. They argued that the environment determines the diversity of social organizations (Crook and Gartlan 1966). This was the first paper to recognize and categorize the diversity of primate social organizations as products of the ecological environment. Relying on the relatively few field studies that had been conducted at that time (De-Vore and Lee 1963), Crook and Gartlan organized primates into five grades (I–V) based on broad categories of gross habitat type and diet (Table 1.1). Although food availability and predation pressure were not listed in their table, greater food seasonality and dispersion, as well as predation pressure, were thought to have selected for the larger groups found in Grades IV and V.

At this early stage, primate socioecology was largely concerned with understanding variation in diel activity, group size, sex ratios within groups, sexual dimorphism, and space use. As the years passed and more species were studied, it became apparent that the Crook and Gartlan model was too simplistic because not all species could be placed into those categories. Nonetheless, the model's logical connection between the environment and social organization was perceived as a major step in the right direction, and future attempts followed this logic. By incorporating other concepts or perspectives, primatologists sought to better explain variation in primate social organization. Below, I highlight

Table 1.1 Selected summary of Crook and Gartlan's (1966) socioecological classification system.

Grade	Habitat	Diet	Diel activity	Group size	Reproductive units	Examples
I	Forest	Mostly insects	Nocturnal	Usually solitary	Pairs	*Microcebus, Cheiro-galeus, Phaner, Daubentonia, Lepilemur, Galago, Aotus*
II	Forest	Fruit or leaves	Crepuscular or diurnal	Very small groups	Small families with one male	*Indri, Lemur, Hapa-lemur, Avahi, Callicebus, Hylobates*
III	Forest-forest fringe	Fruit or fruit and leaves stems, etc.	Diurnal	Small to occasionally large groups	Multi-male groups	*Lemur, Alouatta, Saimiri, Colobus, Cercopithecus, Gorilla*
IV	Forest fringe, tree savannah	Vegetarian-omnivore, occasionally carnivorous	Diurnal	Medium to large groups (*Pan* groups inconstant in size)	Multi-male groups	*Macaca, Presbytis* (now *Semnopithecus*), *Cercopithecus* (now *Chlorocebus*), *Papio cynocephalus, Pan*
V	Grassland or arid savannah	Vegetarian-omnivore, occasionally carnivorous	Diurnal	Medium to large groups	One-male groups	*Erythrocebus, Theropithecus, Papio hamadryas*

what some of the most important contributions added to the development of primate socioecology.

Phylogenetic Constraints as Something Else to Consider (Struhsaker 1969)

Much of the fieldwork on primates in the early years focused on the more easily observed species, and models derived from them suggested a relationship between gross measures of habitat use and both group size and number of adult males in groups. For example, it appeared that small, single-male, multi-female groups were arboreal and lived in forests, whereas large, multi-male, multi-female groups were terrestrial and lived in open habitats. Tom Struhsaker (1969) found that these relationships did not hold among monkeys in West Africa, however. For instance, Preuss's monkeys, *Allochrocebus preussi* (formerly *Cercopithecus l'hoesti preussi*; former names are those used in the cited publications) are terrestrial, and yet they live in single-male, multi-female groups roughly similar in size to those of other closely related arboreal taxa in their tribe (Cercopithecini, commonly known as the guenons). Drills, *Mandrillus leucophaeus*, also do not fit the pattern. They are terrestrial and live in large multi-male, multi-female groups like their relatives, the savannah-dwelling baboons (Tribe Papionini), but they live in closed forests. Extending this to African cercopithecines outside forests, Struhsaker noted that terrestrial, open habitat-dwelling patas monkeys, *Erythrocebus patas*, another member of the Cercopithecini, also live in single-male, multi-female groups. He concluded that, while ecology may be important in determining some aspects of social organization (e.g., solitary males in forested habitats and all-male groups in more open habitats), we might better understand social organizations if we also consider phylogenetic history and constraints (Struhsaker 1969).

Maximizing Reproductive Success, Not Survival (Goss-Custard et al. 1972)

In 1972, John Goss-Custard and two of Crook's students, Robin Dunbar and Pelham Aldrich-Blake, were the first to view primate social organizations as adaptations for maximizing reproductive success rather than simply survival (Goss-Custard et al. 1972). They did not offer a Crook and Gartlan–type model to explain all kinds of social organizations but focused instead on variation in space use and the numbers of males in groups. In particular, they argued that male–male competition is the major determinant of space use (territoriality versus undefended home ranges) and group formation and structure (single-male versus multi-male) under variable conditions of food availability. For example, males would do best to defend territories for "their" females and offspring in order to reduce competition with others for food under fluctuating food supplies, which might occur seasonally or spatially. These authors agreed with Crook and Gartlan (1966) that predation could also select for group living, particularly multi-male, multi-female groups in open habitats where more predators occur and fewer refuges exist compared to forests (Goss-Custard et al. 1972).

Incorporating Intraspecific Variation (Eisenberg et al. 1972)

In that same year, John Eisenberg, Nancy Muckenhirn, and Rudy Rudran (1972) published a model that also focused on male contributions to variation in primate social organizations. Their model recognized five categories of social organization: solitary primates, parental families, uni-male groups, age-graded-male groups, and multi-male groups. The concept of "age-graded-male" groups was introduced to account for a shift from having only one male per group to having multiple males per group over time, which they attributed to a response to greater population densities (Eisenberg et al. 1972). This model incorporated the growing awareness that social organization is not always species-specific and static but, within a given species, can be flexible and responsive to local ecological and demographic conditions (Jay 1968).

Believing that social organizations evolve in a stepwise fashion from simple to complex, Eisenberg and colleagues also laid out two evolutionary pathways to multi-male, multi-female groups, a social organization they considered more "advanced" than other types. From a solitary social organization, one pathway would first lead to family groups; then to uni-male, multi-female groups; and finally, multi-male, multi-female groups. The alternative pathway would first lead to extended uni-male groups, where males are in only periodic contact with groups of females; then to cohesive uni-male, multi-female groups; and then to multi-male, multi-female groups. True multi-male groups were considered an adaptation to predation pressure and were only found in some semi-terrestrial species, such as olive baboons, *Papio anubis*. Finally, group size was viewed as a compromise between competition for limited food resources and the benefits of having others nearby to help locate foods.

While Eisenberg and colleagues assumed the selective force of predation favored multi-male groups among the more terrestrial primates, they suggested that predation and aspects of food resources equally favored single-male groups in arboreal, leaf-eating genera in South America, Asia, and Africa. Nevertheless, they also pointed out that "no single aspect of primate field studies has less supportive data than the generalizations concerning the survival value of the various presumed antipredator mechanisms" (Eisenberg et al. 1972:872).

Predation Takes Center Stage (Alexander 1974)

Richard Alexander, a zoologist with exceedingly broad interests, was interested in, among other things, social behavior as an evolutionarily expressed attribute of individuals. His paper (Alexander 1974) was a precursor to E. O. Wilson's massive book, *Sociobiology: The New Synthesis* (Wilson 1975) in trying to bring natural and social scientists together to view human social behavior from a genetic standpoint. He believed that social behavior requires group living, so he focused on explaining why certain kinds of groups form in the face of conflicts of interest among individuals within groups, which he viewed as necessarily involving trade-offs. He forcefully stated that "There is no automatic or universal benefit from group living. Indeed, the opposite is true: there are automatic and universal detriments, namely, increased intensity of competition for resources, including mates, and increased likelihood of disease and parasite transmission" (Alexander 1974:328). Since groups exist, however, there must have been benefits that outweigh such costs. He entertained three possible benefits for primates living in groups: (1) reduced vulnerability to predation by using the group as cover or for collective defense; (2) access to rare and clumped foods that require others to serve as locators; and (3) access to rare critical resources such as sleeping sites. Alexander ruled out the last two possible benefits because he could find no evidence for primate cooperation in food-finding, and sleeping sites are not sufficiently restricted, leaving him to conclude that predation is "the sole factor capable of causing the (evolutionary) formation and maintenance of primate social groups larger than one or both parents and their offspring. All other aspects of social organization are, in this hypothesis, relegated to a secondary role" (Alexander 1974:332). This focus on predation persists today in most more recent models despite the ongoing paucity of evidence.

Fine-Tuning Food Resource Availability
(Struhsaker and Oates 1975)

As data from field studies steadily accumulated, it was becoming clear that gross dietary categories—such as "fruits" or "leaves" or "insects"— were insufficient to explain primate social organizations. Comparing sympatric eastern black-and-white colobus, *Colobus guereza*, and ashy red colobus, *Piliocolobus tephrosceles* (formerly *Colobus badius*), Tom Struhsaker and John Oates (1975) suggested that differences in social organization are determined more by variation in food distribution. In the Kibale Forest Reserve (now National Park), black-and-white colobus live in small groups of 8–15 individuals, typically including just one adult male, whereas red colobus live in larger groups of 19–80 individuals, including multiple males. Struhsaker and Oates (1975) pointed out that, although both colobus species feed heavily on leaves, black-and-white colobus obtain their food from a few food sources within small areas, which they argued allows small groups to live within small, defensible home ranges. In contrast, the much more diverse diet of red colobus requires that they obtain their foods within larger areas, making home ranges difficult to defend and enabling multiple males to join the groups (Struhsaker and Oates 1975). Predation was not discussed, perhaps because the close proximity of their home ranges suggested the two study groups were under similar predation pressure.

The Arrival of Statistical Meta-Analyses
(Jorde and Spuhler 1974; Clutton-Brock and Harvey 1977a)

The first quantitative meta-analysis to examine primate social organizations using statistical approaches was published around the same time (Jorde and Spuhler 1974). Analyzing data for 19 variables involving ecology, demography, and social behavior, and drawing largely from catarrhine primates (monkeys and apes in Africa and Asia), anthropological geneticists Lynn Jorde and James Spuhler found, among other results, a positive correlation between female body mass and home range size. Their principal components analysis identified four clusters that closely aligned with Crook and Gartlan's Grades II–V (Grade I was not represented as

7

no nocturnal strepsirrhines—lemurs, lorises, and galagos—were included in their analyses), which they interpreted as a sort of confirmatory check on their analyses.

In their own quantitative analyses of body mass, feeding group mass and size, home range size, day range length, population density, biomass, sex ratio, and sexual dimorphism, Tim Clutton-Brock and Paul Harvey (1977a) confirmed other qualitative generalizations, including that terrestrial primates tend to live in larger groups and have larger home ranges than arboreal primates and that nocturnal primates tend to be small and insectivorous. Their quantitative analyses also identified clear patterns that established a consensus on some socioecological debates, e.g., that sexual dimorphism is better viewed as a consequence of sexual selection, not predation pressure.

They classified primates into seven types based on diel activity, substrate use, and gross diet but not on group size or numbers of males in groups, while noting that species in the same genus and the same category tended to have similar social organizations:

1. nocturnal arboreal insectivores;

2. nocturnal arboreal frugivores;

3. nocturnal arboreal folivores;

4. diurnal arboreal frugivores;

5. diurnal arboreal folivores;

6. diurnal terrestrial frugivores; and

7. diurnal terrestrial folivores.

On the question of variation in social organizations, Clutton-Brock and Harvey suggested that it might be more useful to recognize multiple selective pressures, including both food and predation, and to focus on constraints that might limit group size, such as reduced feeding rates and becoming more conspicuous to predators. However, they leaned more heavily toward predation as an explanation for small group size in nocturnal primates and large group size in terrestrial primates, with food distribution as secondary. Throughout their paper, they bolstered their findings by drawing parallels with other taxa, especially birds and ungulates (Clutton-Brock and Harvey 1977a).

Shift in Focus to Ecological Determinants of Female Groups and Social Relationships (Wrangham 1980)

Predation was, however, relegated to the backstage in Richard Wrangham's (1980) influential paper on the evolution of group living among females. In proposing a model to explain the evolution of multi-female groups, he argued that intergroup competition for food was the driving force behind female groups and, further, that food distribution determines the expression of female social relationships within and between groups. Since it is assumed that females are most limited in their reproductive success by access to food, he argued that groups would form as the outcome of coalitionary competition for high-quality, clumped food resources and that larger groups would outcompete smaller groups. These coalitions would be composed of female relatives because kin would gain the most from helping each other. Consequently, females would remain in their natal groups throughout life and form stable dominance hierarchies within groups. Wrangham labeled these species "female-bonded." They were contrasted with females that feed on undefendable foods, i.e., low-quality, widely distributed foods, and thus have only weakly expressed dominance hierarchies within groups, disperse readily, and form groups around males for their protection ("non-female-bonded groups") (Wrangham 1980). This model effectively shifted attention away from the effort to understand basic differences in primate social organization, i.e., solitary vs. group living, group size and composition, and diurnal versus nocturnal living, and more toward understanding social relationships among females within and between multi-female groups—a trend that has persisted to the present day.

Predation Returns to the Limelight (van Schaik 1983)

Three years later, Carel van Schaik (1983) offered an analysis that he argued did not support Wrangham's intergroup competition hypothesis and, therefore, supported Alexander's view of the primacy of predation pressure as the selective force behind group living in primates. Examining data from 27 populations of 14 species, he found fewer infants per female in larger groups, which he interpreted as an indication of greater food competition within than among groups. Unfortunately, it later became clear that eight (57%) of the primate species included in his analysis (Verreaux's sifakas, *Propithecus verreauxi*; chacma baboons, *Papio ursinus*; black-and-white colobus; red howler monkeys, *Alouatta arctoidea* (formerly *A. seniculus*); mantled howler monkeys, *Alouatta palliata*; black-crested Sumatran langurs, *Presbytis melalophos*; Hanuman langurs, *Semnopithecus entellus*; and Nilgiri langurs, *S. johnii*—the latter two species formerly in the genus *Presbytis*), representing 16 (59%) of the populations in his analysis, are now known to experience infanticide (Rowe and Myers 2016). As larger groups are often more attractive to infanticidal males (Crockett and Janson 2000; Steenbeek and van Schaik 2001), infanticide, rather than food competition, could account for his findings. Despite not testing the predation hypothesis with any data on predation (remember Eisenberg et al. 1972?), this paper solidified the prominent position of predation in favoring multi-female groups. Today, most primatologists consider predation to be the main selective pressure responsible for grouping itself, whereas food competition is viewed as determining both the upper limit on the size of groups and the nature of female social relationships within and between groups (van Schaik 1983; Janson and Goldsmith 1995; Sterck et al. 1997).

More Iterations on Female Relationships

Four, Not Two, Types of Females (van Schaik 1989; Sterck et al. 1997)

Predation was again emphasized as the main selective pressure favoring group living in a model proposed by van Schaik (1989) that he later modified with Liesbeth Sterck and David Watts (Sterck et al. 1997) to bring infanticide in as a selective pressure favoring female philopatry, which they defined as staying in one's natal group. In these models, food competition was viewed as an unavoidable consequence of grouping in response to predation pressure, as Alexander (1974) claimed. They extended it by asserting that, because females in groups are spatially closer to each other than they are to females in other groups, food competition will always be stronger within groups than between groups. In these two models, four types of females emerge as "competitive regimes" primarily from variation in predation and food distribution: Type A or Dispersal-egalitarian, which both correspond to Wrangham's non-female-bonded classification but have high predation; Type B or Resident-nepotistic, which both correspond to Wrangham's female-bonded classification but have high predation; Type C or Resident-egalitarian, which both have weakly expressed dominance hierarchies, female philopatry, and intermediate to high predation; and either Type D (in van Schaik's 1989 model), which has weakly expressed dominance hierarchies, female philopatry, and low predation, or Resident-nepotistic-tolerant (in Sterck and colleagues' 1997 model), which has highly expressed dominance hierarchies, female philopatry, and low predation. The predation-dependent model by Sterck and colleagues appears to be the most commonly accepted one at this time.

Three, Not Two or Four, Types of Females (Isbell 1991)

Competition for food can be direct via contest, which is expressed as fighting or the threat of fighting. It can also be indirect via scramble, which occurs when animals reduce foods available for others by consuming the food first (Janson and van Schaik 1988), as might occur when group home ranges overlap, for example. More than one group feeding in the same area results in lower food abundance, which forces groups to search more widely for food (intergroup scramble competition) (Isbell 1991). Scramble competition also occurs when food patches get depleted more quickly by larger groups, causing individuals to incur higher energetic costs from needing to travel farther per day (intragroup scramble competition) (Waser 1977). I argued that these types of competition could be expressed in a variety of combinations reflecting scramble and contest competition within and between groups (Isbell 1991). My analysis showed that the guenons (*Cercopithecus* spp.) do not fit neatly into Wrangham's categories as either female-bonded or non-female-bonded. Although they could potentially fit into van Schaik's Type C or Type D categories, in my model—as with Wrangham's—there was no need to invoke predation as a selective force in social relationships. Although female guenons remain in their natal groups throughout life and are aggressive toward other groups, available evidence suggested that their dominance hierarchies were only weakly expressed. I attributed their combination of contest competition between groups and little or no contest competition within groups to limited food abundance and widely dispersed foods, respectively. By spreading out as they feed, I suggested that guenons would be able to avoid the foraging paths of others and the energetic cost of traveling farther per day in larger groups, thereby minimizing contest and scramble competition within groups (Isbell 1991). This type of female would not necessarily suffer increased food competition by living in larger groups, in contrast to Alexander's and van Schaik's models.

Pushback against Socioecology
(Di Fiore and Rendall 1994; Thierry 2008)

In 1994, Tony Di Fiore and Drew Rendall, then graduate students at UC Davis, examined female social relationships from a phylogenetic perspective (Di Fiore and Rendall 1994). Harkening back to Struhsaker's discussion of guenon social organization as an example of phylogenetic constraints (Struhsaker 1969), they concluded that social organizations and relationships are conservative, with qualities that remain stable within lineages despite sometimes dramatic changes in habitat. For instance, the many species of macaques all live in large groups of females with clear dominance hierarchies and female philopatry despite their wide biogeographical range across habitats that surely must be variable in food distribution and abundance (Di Fiore and Rendall 1994).

Their challenge of the primacy of ecological determinants of social organizations and female social relationships was later reinforced by Bernard Thierry (2008), who pointed out the many inconsistencies of the Sterck et al. (1997) model in predicting details of social relationships among females, e.g., variation in food patchiness does not affect the rate of agonism in the predicted direction for some species. With the shift in socioecology's focus away from explaining basic phenomena such as variation in social organizations and toward explaining detailed variation in female social relationships, Thierry declared that socioecology was dead and that it was time to move on to other approaches. He suggested that one such approach might be to conduct phylogenetic analyses on a variety of social behaviors, such as measures of dominance and conflict management, to determine which ones have either strong or weak phylogenetic signals and which ones are linked together within a particular lineage (Balasubramaniam et al. 2012a, b; Thierry 2013).

Where We Are Now

The search for ecological explanations for the diversity of social organizations was unfinished before it was co-opted at the turn of the century by an ultimately unsatisfactory focus on socioecological explanations for variation in female social relationships. At the same time, the development of non-invasive techniques to study physiology, endocrinology, and genetic relationships were revealing answers to other long-standing challenges, such as clarification of paternity. It is not surprising, then, that the attention of fieldworkers has been drawn to other more proximate questions. Without a strong socioecological model and yet unable to imagine ecology not affecting primate social organizations (if not social relationships), we now seem to be content with repeating the view that predation favored group living in diurnal primates (as well as solitary foraging in nocturnal primates) while intragroup food competition puts a cap on group size, as if these assertions have been well-tested and confirmed. Some studies do report a decline in reproductive success in the largest groups (e.g., blue monkeys, *Cercopithecus mitis*; Roberts and Cords 2013), but alternative explanations are often not considered. For instance, the same result could occur if intergroup competition prevents groups from expanding their home ranges to accommodate additional group members. Perhaps we no longer look for alternatives because that view has become so ingrained in the discipline.

Where I Want to Take You

I want to take you back to socioecology before female social relationships became its focus. I want to take you away from the currently accepted socioecological view that predation and food availability (distribution and abundance) are the primary selective pressures shaping primate social organizations. What is responsible for variation in primate social organizations? Nearly 60 years after that question was first raised, our emphasis on predation and food availability has not given us a good answer. It is high time for closure (or at least a new open door), but the status quo is not getting us there.

To begin, let us acknowledge that so much more is known about how primates live than back in 1966, and it has allowed us to put together a more complete classification scheme of the diversity of primate social organizations. Social organizations have been defined in different ways over the years, but the definition provided by Peter Kappeler and Carel van Schaik (2002:709)—"the size, sexual composition, and spatiotemporal cohesion of a society or social unit"—fits my purpose best because it is closest to what Crook and Gartlan had in mind. At the most basic level, the following social organizations commonly recognized at present are organized around spatial and social conditions but without a functional (explained by natural selection) component:

1. solitary foragers: adults travel about in their own home ranges, which may or may not overlap with other home ranges;

2. pair-living animals: one adult of each sex shares the same home range, and they may or may not travel with each other;

3. single-male, multi-female groups: multiple breeding females travel together along with one adult male in the same home range;

4. multi-male, multi-female groups: multiple breeding females travel together along with more than one adult male in the same home range; and

5. fission-fusion communities: adult females travel largely independently of one another in a shared home range that typically includes multiple adult males.

I hope to convince you that primate social organizations will be better understood if they are classified functionally in the context of ecology. It starts out by examining space use by females and then examines space use by males. Variation in female space use creates female social organizations, and then male space use, when layered onto female space use, completes primate social organizations. Please note that I am not concerned here with mating systems, which have sometimes been conflated with social organizations (Alexander 1974; Müller and Thalmann 2000). For example, pair living, which describes spatial behavior, is sometimes equated with monogamy, which describes sexual behavior, because in both cases, one adult male and one adult female are involved. Here I am concerned with the use of space by individuals because it is what leads to our perception of different types of social organizations. I hope to convince you that there is a unitary explanation for variation in primate social organizations, something many socioecologists have long given up trying to find.

Specifically, I hope to convince you in the following pages that predation does not hold up well as the main selective pressure driving the evolution of different primate social organizations. Instead, I will propose that the main selective pressure is foraging efficiency, with greater foraging efficiency meaning greater returns in the currency most important to the animal, such as calories, avoidance of secondary compounds, or mixing nutrients (reviewed in Felton et al. 2009), for the effort invested in the process of acquiring food. I will further propose that foraging efficiency is tempered by the relative importance of needing certainty in knowing the location of foods. Foraging efficiency will be affected by how successfully primates that require food location certainty find food at their expected locations and by how successfully primates that do not require food location certainty find food as they move along. Foraging efficiency will be poorer for those who rely on food location certainty if they arrive to find an absence of food where it was expected. Similarly, foraging efficiency will be poorer for those who do not rely on food location certainty if, as they move along, they find less food on one day than another for the energy they expended. Before I begin my efforts to convince you that there is a better way forward though, it is important to provide you with a bit of background on dispersal, because it is intimately associated with food location certainty and differential space use, links that ultimately affect social organizations.

CHAPTER 2

Dispersal and Philopatry

Female settlement and ranging patterns . . . are fundamental to the evolution of the social system.

— J. M. Williams et al. (2002:347)

Natal dispersal, the movement of individuals of one or both sexes away
from the mother (social dispersal) or birthplace (locational dispersal) is
universal in mammals, including primates (Greenwood 1980; Pusey and
Packer 1987; Isbell and Van Vuren 1996). The fact that it is so widespread
tells us that there has been strong selection for it; however, as long-term
research on primates has shown, there is wide variation in who leaves,
where they go, and what conditions encourage their leaving (Jack and
Isbell 2009). Understanding dispersal and its opposite, philopatry, is crucial to understanding the diversity of primate social organizations. Any
socioecological model worth its salt should account for variability in the
extent of locational dispersal, i.e., how far individuals go from their natal home range, if they go at all.

The distinction between social and locational dispersal may not
matter much when considering solitarily ranging animals if, when they
move away from the mother, they also move away from the mother's
home range. In such cases, dispersal involves both social and spatial dis-
placement. The distinction is more important if dispersers, as they

establish their new home range, continue to overlap with a portion of the mother's home range, because social and locational dispersal can then become unlinked. In such cases, mother and offspring may avoid interacting or continue to interact with each other in the area of overlap. The offspring can also choose to spend more time in familiar or unfamiliar parts of its new home range. These various combinations of locational and social dispersal are perhaps more obvious with group-living animals. If a group-living individual disperses to a group with no home range overlap with the natal group, both social and locational dispersal occur. If that individual disperses instead to a group whose home range overlaps extensively with that of the natal group, it disperses socially but only a little locationally. Finally, it is possible to experience locational but not social dispersal in unusual situations. Yellow baboons, *Papio cynocephalus*, facing a deteriorating habitat, and vervets, *Chlorocebus pygerythrus* (formerly *Cercopithecus aethiops*), facing both a deteriorating habitat and intergroup competition, dispersed locationally but not socially when groups moved into areas where the adult females and their offspring had never been before (Isbell et al. 1990; Altmann and Alberts 2003). Vervets did this in two ways—when entire groups shifted their territories and displaced neighboring groups as a result of the natural loss of many of their trees (Isbell et al. 1990), and when groups became so small that the remaining members abruptly abandoned their territories and joined neighboring groups (Isbell et al. 1991).

The distinction between social and locational dispersal is important to make because there are different costs associated with the two types of dispersal, and those costs can affect dispersal decisions of individuals. It may be easier for animals to disperse socially if they do not also disperse locationally because there are fewer potential costs, for instance. One potential cost of social dispersal is aggression from strangers, which may depress reproductive success through aggression-induced stress, poorer access to foods, or even death. Social dispersers may also be more vulnerable to predators if they disperse alone and have no companions to help with predator detection. Potential costs of locational dispersal include greater vulnerability to predators because it takes time to learn about locations of refuges, effective escape routes, and the habits of individual predators. Locational dispersers might also have more difficulty finding food, but as I will argue in later chapters, there may be substantial variation in the severity of this cost, depending on how primates move about to get their food.

Studying dispersing animals is challenging, and documenting the costs of their actions is even more challenging. Nonetheless, there are some good examples showing costs of dispersal. The groups of vervets that shifted their territories in response to habitat loss experienced more predation, mostly from leopards, *Panthera pardus*, when they spent more time in new areas (Isbell 1990; Isbell et al. 1990). Similarly, members of remnant groups that abandoned their home ranges and fused with neighboring groups had higher mortality rates, again most likely from leopard predation, than members of their adopted groups during their first six months together. After going through that "learning curve," their mortality dropped to that of the groups they joined (Isbell et al. 1993). These studies show that predation-driven mortality can indeed be a cost of locational dispersal.

Female chimpanzees, *Pan troglodytes*, who dispersed both socially and locationally in Gombe National Park, Tanzania, reproduced for the first time about two years after females who remained in their natal community (Walker et al. 2018). Perhaps their delayed reproduction represents a cost of social dispersal because they experienced high levels of aggression from resident females in the new community, resulting in physiological stress and low social status (Kahlenberg et al. 2008; Pusey et al. 2008). However, their delayed reproduction could also be a cost of locational dispersal. The effects of unfamiliarity with the new home range on food acquisition have not been reported to my knowledge.

Male gray-cheeked mangabeys, *Lophocebus* (formerly *Cercocebus*) *albigena*, in Kibale National Park, Uganda, often travel far (≥ 200 m) from their groups even when not in the process of dispersing, and when they move away temporarily, their risks may be similar to those experienced by dispersing males because in both instances, they are alone (Olupot and Waser 2005). This suggests that some of the costs of dispersal are social in origin for them. Males who were alone or in the process of dispersing (both called isolated) appeared to adjust to their risks by seeking out red colobus groups. They also scanned more often than males closer to groups, perhaps perceiving greater vulnerability to aggressive conspecifics and predators since encounters with other isolated males elicited mutual agonism, and mortality was higher for isolated males than for males near or within groups. Two of five isolated males possibly died from falls, either during aggressive encounters with other males or from encounters with predators, whereas three were confirmed to have died of predation, most likely from crowned eagles, *Stephanoaetus coronatus* (Olupot and Waser 2005). Unfortunately, it was not reported if they died in areas unfamiliar to them. They did not appear to suffer reduced foraging efficiency, because isolated and group-associated males had similar percentages of time spent feeding, foraging, and moving, and they maintained similar travel speeds (Olupot and Waser 2001, 2005). One would not expect their foraging efficiency to decline if they remained in familiar areas. Again, however, time spent in familiar versus unfamiliar areas was not reported.

In a later study of that same population however, an entire group moved into an unfamiliar area, allowing Karline Janmaat and Rebecca Chancellor (2010) to examine the group's foraging efficiency, with the expectation that it would decrease in unfamiliar areas. They found no significant difference in the number of fruit trees fed in or the time spent feeding on fruit between unfamiliar and familiar areas for either adult males or adult females. However, adult males were less efficient in unfamiliar areas at finding individual trees of one particular fig species they preferred. This did not also hold for adult females, perhaps because they rarely fed on that species in either area. The group also appeared to invest more energy in searching for foods in unfamiliar areas because they traveled farther per day in new areas, but they also moved in wider swaths and so covered more area, which could have compensated for the greater daily travel distances by making it easier to discover foods, if they use each other to locate foods.

The clearest example of how high the separate costs of social and locational dispersal can be, especially when animals are forced to disperse from their natal group and home range, comes from female red howler monkeys. Social dispersal was associated with receiving more aggression and injuries than group females, who are aggressive toward extragroup females (Sekulic 1982a; Crockett and Pope 1988). Locational dispersal was associated with a poorer diet—including more ash-free neutral detergent fiber, which is indigestible, and less protein and phosphorus—compared to residents of established groups (Pope 1989). These and other costs, such as high rates of male replacement in groups newly formed by dispersers, leading to more frequent infanticide and group dissolution, conspire to reduce reproductive success among dispersing females compared to females able to remain in the group (Crockett and Pope 1993).

Clearly, dispersal is potentially very costly for both males and females, and yet it has been widely favored by natural selection, which means that the one benefit that must ultimately outweigh the potential costs of dispersal is greater reproductive success, on average, for dispersers compared to what they would have achieved had they remained in their natal home range with familiar conspecifics. For the average dispersing male, it seems straightforward that going elsewhere will increase his access to more unrelated females and thus potentially more breeding partners and greater reproductive success. For females, leaving the natal group on one's own accord might be more expected when their potential costs of dispersal are low to begin with, e.g., when strangers do not react aggressively toward them, and they can remain in familiar areas where they know about the food and predators. In such cases, other factors that affect reproductive success may rise in importance in dispersal decisions. For instance, females might disperse to avoid infanticidal males (e.g., mountain gorillas, *Gorilla beringei*, formerly *G. gorilla*; Robbins et al. 2009a), to avoid mating with their fathers or other male relatives (e.g., Milne-Edwards' sifakas, *Propithecus edwardsi*; Morelli et al. 2009), or to join more or higher-quality males for protection against infanticidal males or predators (e.g., red colobus; Marsh 1979; Struhsaker 2010). Irrespective of the influence of male behavior on female dispersal decisions, females might disperse again (called secondary dispersal) after reproduction fails because of illness; energetic, physiological, or hormonal insufficiency; or the death of their infant for reasons other than infanticide (Isbell and Van Vuren 1996; Isbell 2004). Female mountain gorillas, for example, are more likely to transfer to new groups after their infant has died (Robbins et al. 2009b). In short, there can be multiple beneficial reasons for female dispersal when social and locational costs of dispersal are low.

Do Existing Categories of Social Organizations Predict Dispersal Patterns?

The distinction between social and locational dispersal is important to recognize not only because these forms of dispersal are associated with different costs, but also because dispersal involves space use and dispersal patterns are often thought to be consistent across populations within each of the current commonly recognized categories of primate social organizations. The examples I provide below will highlight that there can be more variation in dispersal patterns within some of these types of social organizations than we currently acknowledge, which suggests that we are forcing some primates to fit into categories that may be more arbitrary than functional.

Variation in Dispersal Patterns
Solitary Foragers

Primates that move about their home ranges by themselves as they forage for food disperse from their natal home ranges both socially and locationally to varying degrees. Gray mouse lemurs, *Microcebus murinus*, in the Reserve Forestière d'Ampijoroa of northwestern Madagascar exhibit social and locational dispersal patterns typical of solitarily foraging mammals in general: natal males travel farther than females from the mother's home range (Waser and Jones 1983; Radespiel et al. 2003). In one study, median natal dispersal distance was 251 m for males, with more than half of the males dispersing more than one home range diameter away from their probable mothers. In their first breeding season, adult males only stayed nearby if close female relatives had died or disappeared (Radespiel et al. 2003). In contrast, median natal dispersal distance was 63 m for females, and 86% dispersed one home range diameter or less from their probable mothers. More than half of the females remained near their presumed mothers during their first breeding season (Radespiel et al. 2003). Female home ranges overlap with, on average, 1.5 other females (Radespiel 2000). The extent of home range overlap is greater among females that form sleeping groups during the day (Radespiel 2000), and females that sleep together are more closely related than those that do not (Radespiel et al. 2001). It is likely that many sleeping groups include mothers and adult daughters. By establishing home ranges that overlap with their natal home ranges, females only partially disperse both locationally and socially.

Although dispersal among solitarily foraging pale fork-marked lemurs, *Phaner pallescens* (formerly *P. furcifer*), has not been observed directly, in Kirindy Forest, Madagascar, they must normally disperse both socially and locationally because home range overlap is low. Female home ranges overlap by only about 11% (Schülke and Kappeler 2003), suggesting that daughters disperse away from their natal home ranges and mothers more than do gray mouse lemurs. Immigrant males have been observed taking over the home ranges of resident males that have died or disappeared, and a female did not disperse but remained on her natal home range when her presumed mother died (Schülke 2003). Unlike mouse lemurs, some natal individuals delay their dispersal until several years after they have become adults (Schülke 2003). Although fork-marked lemurs forage solitarily, their dispersal patterns appear more like those of cohesive pair-living primate species, and indeed, the home range of a given male will overlap almost entirely with that of just one female (Schülke and Kappeler 2003). These two species provide good examples of variable dispersal patterns, i.e., the extent of social and locational dispersal and the timing of dispersal, within the category of solitary foragers.

Diurnal Pair-Living Primates

In pair-living primates, both sexes disperse but, again, with varying degrees of social and locational dispersal depending on the dispersal distance and extent of home range overlap. Like some nocturnal fork-marked lemurs, diurnally active, white-handed gibbons, *Hylobates lar*, in Khao Yai National Park, Thailand, delay their dispersal until well after they reach adult size. The rarity of floaters—dispersers not yet settled on a home range—also suggests that they delay dispersal (Brockelman et al. 1998). Dispersers can move directly to a new home range by replacing the same-sex resident, filling a vacancy due to the resident's death, finding vacant land, or, in the case of males, joining as the second male (Brockelman et al. 1998). Unlike fork-marked lemurs however, home ranges of white-handed gibbons in this population overlap considerably (Brockelman et al. 1998). Genetic analyses confirm behavioral observations that males more often disperse to adjacent home ranges where male relatives sometimes reside, whereas females more often disperse to nonadjacent home ranges (Matsudaira et al. 2017). Thus, in general, females disperse both socially and locationally more than do males.

In contrast to white-handed gibbons, natal male and female Azara's owl monkeys, *Aotus azarae*, in the Gran Chaco of Argentina do not typically delay dispersal but leave on their own volition as they become adults, or they disperse after an immigrant evicts one of their parents (Fernandez-Duque 2009). They then become floaters as they search for a home range without a same-sex adult or for one that they can take over (Huck and Fernandez-Duque 2017). There is no apparent sex difference in dispersal distance, and most dispersers settle in home ranges nonadjacent to the natal home range (Fernandez-Duque 2009; Wartmann et al. 2014). Thus, natal dispersal appears to be both social and locational for most Azara's owl monkeys, regardless of sex. These examples again show variation in the timing and extent of social and locational dispersal within one type of social organization.

Single-Male, Multi-Female Groups

One might expect male dispersal and female philopatry in single-male, multi-female groups given that the birth sex ratio in groups is more even than the adult sex ratio. Indeed, blue monkeys in Kakamega Forest, Kenya, meet these expectations. The number of females in Kakamega blue monkey groups ranges from 10 to 23. While there may be temporary influxes of males into groups during the mating season and supernumerary males in the largest groups, the modal pattern is that only one adult male remains resident with the group the rest of the year

(Ekernas and Cords 2007). Males do not delay dispersal but usually leave their natal groups voluntarily a year or so before reaching adulthood (Bronikowski et al. 2016), not because of increased aggression or reproductive seasonality but in relation to fruit availability (Ekernas and Cords 2007). During the dispersal process, males rarely join another group directly but more commonly disappear and reappear in their natal groups for up to about five weeks before completing the dispersal transition (Ekernas and Cords 2007). Sometimes males go quite far away, leaving the study area and then returning months later (Ekernas and Cords 2007). This temporary nomadism presumably helps them become familiar with areas outside their natal home range. If males eventually settle in groups adjacent to their natal home ranges, they will certainly be familiar with some of their new group's home range because, although blue monkeys are territorial, their home ranges overlap by as much as 50% (Cords 2007). Thus, all male blue monkeys disperse socially but the extent of locational dispersal varies depending on the individual.

Male dispersal and female philopatry in single-male, multi-female groups may just be associated with large multi-female groups, however. Milne-Edwards' sifakas in Ranomafana National Park, Madagascar, live in minimally overlapping home ranges in small groups that include, on average, only one or two reproductively active females and one reproductively active male (Morelli et al. 2009; Gerber et al. 2012). Such small multi-female groups suggest that both males and females disperse, and indeed, they do. Most delay dispersal and remain in their natal groups as nonreproductive adults for up to five years after reaching sexual maturity before they finally do disperse. Exceptionally, if the opposite-sex parent is replaced by an immigrant, individuals will reproduce in their natal group before dispersing (Morelli et al. 2009). The peak time for dispersal is in the three months before the breeding season (Morelli et al. 2009). Natal males and females appear to disperse to find unrelated mating partners (Morelli et al. 2009); hence, dispersal for both sexes is social and locational.

Red howler monkeys in Venezuela live in slightly larger multi-female groups than Milne-Edwards' sifakas but never with more than four adult, reproductively active females (Crockett 1984). Natal dispersal of females depends on the number of reproductively active females already in the natal group, increasing in likelihood as the group reaches its maximum number (Pope 2000). Dispersal is forced upon females in larger groups before they reach adulthood by aggression from adult females who are not their mothers. They do not immigrate into existing groups but float until they can create their own group by defending a territory with other unrelated and groupless females and males (Pope 2000). Males tend to stay longer than females in the natal group, helping their father defend the territory. When they leave, they stay closer to their natal home range than do females, sometimes even continuing to stay in much of their natal home range as solitaries (Pope 2000). Thus, while females immediately disperse both socially and locationally, males disperse socially and eventually locationally provided they can find a group and territory of their own.

Groups of Thomas's langurs, *Presbytis thomasi*, near the Ketambe Research Station, Sumatra, live in extensively overlapping home ranges and may have up to seven adult females along with the one adult male (Steenbeek and van Schaik 2001; Sterck et al. 2005). Natal females disperse when the resident male is their father, forming groups by joining a different adult male and continuing to disperse in later years, resulting in fluid group membership (Sterck 1997; Steenbeek and van Schaik 2001; Sterck et al. 2005). Natal juvenile males disperse with or without aggression from the resident adult male to live solitarily, or they may join all-male bands while continuing to mature (Sterck 1997). Dispersing females appear to stay closer than dispersing males to their former home ranges, as all-male bands are more fluid than bisexual groups in their ranging behavior (Sterck 1997). Thus, females disperse socially but not always locationally, whereas males do both (Sterck 1997).

These four examples highlight that, even in single-male, multi-female groups, dispersal patterns are variable. With only one male per group, we might reasonably infer that, in the absence of direct evidence (recall that dispersal is difficult to study), natal males disperse, at least socially, especially if the father remains. Female dispersal and philopatry decisions are more variable, however, and we would not be able to infer anything about their dispersal patterns without direct evidence.

Multi-Male, Multi-Female Groups

Inferring dispersal patterns in multi-male, multi-female groups based on the adult sex ratio or female group size is even more challenging. Gray-cheeked mangabeys in Kibale live in groups of up to 10 adult females and 10 adult males (Olupot and Waser 2005). Like other cercopithecines, females are known to remain in their natal group throughout their lives, whereas males disperse. Dispersal is a solitary undertaking for them. Males begin to disperse from their natal groups in subadulthood, a process that can take many months (Olupot and Waser 2005). They are solitary for much of this time but sometimes visit other groups. Initially, these visits elicit aggressive responses by resident males. With persistence and mating opportunities however, dispersing males can become new group members. While natal males disperse socially, they can minimize locational dispersal by immigrating to an adjacent group whose home range overlaps with that of the natal group; males can also move far-ther away, as shown by two radio-collared males who crossed at least two home ranges to join new groups (Olupot and Waser 2001, 2005).

As cercopithecines, vervets also exhibit male dispersal and female philopatry. Groups of 2–10 adult females and 2–7 males (Cheney and Seyfarth 1981; Isbell and Jaffe 2013) typically defend territories, but there is also some home range overlap (Cheney 1981; Isbell et al. 1990, 2002, 2021). Near Mpala Research Centre, Laikipia, Kenya, home ranges over-lapped with any one other home range by 13–35% because neighbors made incursions when the other group was not nearby (Isbell et al. 2021). Unlike gray-cheeked mangabeys, in Amboseli (and almost cer-tainly Segera Ranch, Kenya, given the limited dispersal options available there), natal males spend no time alone and typically disperse, some-times with other males, to adjacent groups where previous group mem-bers had gone before (Cheney and Seyfarth 1983; Isbell et al. 2002; see also Young et al. 2019). Under such conditions, the extent of social and locational dispersal for natal male vervets appears to be less than ex-pected for a territorial, multi-male, multi-female species.

Northern muriquis, *Brachyteles hypoxanthus* (formerly *B. arachnoi-des*), at the Estação Biológica de Caratinga, Minas Gerais, Brazil, live in groups of 8–19 adult females and 6–14 adult males (Dias and Strier

2003). Males remain in their natal group for life, but dispersal is the norm for females, who leave their natal group before sexual maturity. Hormonal profiles reveal that they do not disperse in response to reproductive suppression, as is the case for some callitrichines (marmosets and tamarins) (Strier and Ziegler 2000). They may visit other groups temporarily before permanently dispersing to their new group where they then begin reproductively cycling, although they typically reproduce only after three years in their adopted group (Strier et al. 2015). Females do not move preferentially to groups with maternal relatives or other previous members of their natal group, thus apparently dispersing socially (Strier et al. 2015). With no more than 25% overlap between home ranges (Lima et al. 2019), some degree of locational dispersal occurs when females join another group. Some females have dispersed from their natal home range into new areas uninhabited by other muriqui groups, however (Strier et al. 2015), confirming that full locational dispersal can also occur.

Red colobus monkey groups in Kibale include at least 2–22 adult females and 2–16 adult males (Struhsaker 1975:223–225; Isbell 1983, 2012). Although both males and females can disperse from their natal groups, females do so more often. Because groups have extensively overlapping home ranges (Struhsaker 1975:17), dispersal appears to be social but only minimally locational. Dispersing males initially tend to range solitarily, whereas females transfer directly to new groups as juveniles, often after visiting neighboring groups during intergroup encounters (Struhsaker 1975:16, 2010:108). Females can easily sample groups because intergroup encounters are frequent. In two studies, focal groups were within 50 m of at least one other group on 49% (Struhsaker 2010:92) and 59% (Isbell 1983) of all observation days. Females apparently prefer to join a group and stay in it longer when the group is growing than when it is declining (Struhsaker 2010:110), which is consistent with the finding that genetic relatedness among females was weaker in a large group than in a small group in that population (Miyamoto et al. 2013), as larger groups would attract more females. It is intriguing that female red colobus would prefer groups that are growing since food competition has been argued to be more intense in larger groups (Snaith and Chapman 2008; Gogarten et al. 2015; but see Isbell 2012). We will return to the topic of female red colobus and what limits their reproductive success in a later chapter. In any case, these examples again reveal that dispersal patterns are more variable than might be expected in multi-male, multi-female groups.

Fission-Fusion Communities

Chimpanzees and bonobos, *Pan paniscus*, both exhibit male philopatry and female dispersal, although some female chimpanzees do remain in their natal group (often referred to as a "community"). Natal female chimpanzees disperse approximately one year after sexual maturity to join a neighboring community (Boesch and Boesch-Achermann 2000:47; Nishida et al. 2003; Walker et al. 2018), whereas natal female bonobos disperse at puberty (Gerloff et al. 1999; Toda et al. 2022) and often spend time with neighboring communities encountered in areas of home range overlap before finally settling down with one (Sakamaki et al. 2015). Although females disperse socially in both species, they apparently differ in the extent of locational dispersal. Chimpanzee community home ranges overlap only minimally (Herbinger et al. 2001; Wilson et al. 2004), but there appears to be considerable overlap between bonobo community home ranges (Sakamaki et al. 2018; Lucchesi et al. 2020), suggesting that female chimpanzees disperse locationally to a greater extent than female bonobos.

Main Points

The ubiquity of dispersal makes understanding its proximate and ultimate causes, and its consequences, of wide interest. Natal social and locational dispersal are particularly relevant for understanding social organizations; however, as the examples reveal, dispersal patterns are variable within our current classification of social organizations, which is based on numbers of adult males and females sharing a home range. Philopatry versus dispersal; delayed dispersal versus dispersal before, at, or after sexual maturity; floating versus direct movement into new groups; and settlement in familiar versus unfamiliar home ranges cannot be easily predicted by our existing classification system but may be more related to costs of social and locational dispersal for individuals. Since dispersal (or the lack thereof) is at the heart of settlement patterns, and settlement patterns are core to understanding social organizations, classifications of social organizations would be more useful if they take into account dispersal behavior.

CHAPTER 3

The Variable Home Range Sharing Model and Its Classification System for Primate Social Organizations

> To investigate the degree of range overlap is especially relevant for understanding the evolution of primate social systems.
>
> —F. M. Wartmann et al. (2014:921)

Ultimately, social organizations are about space use. Individuals need access to an area, a home range, that satisfies their needs. We know that (1) across primates, group-living species with heavier group biomasses almost always have larger home ranges than those with lighter biomasses (Milton and May 1976; Clutton-Brock and Harvey 1977a); (2) within a given species, larger groups tend to have larger home ranges than smaller groups, all else being equal (Clutton-Brock and Harvey 1977a; Isbell 1991; Campos et al. 2015); and (3) home ranges tend to be larger in poorer-quality habitats (Struhsaker 1967; Riley 2008). I think we can safely infer that the size of home ranges at least reflects food abundance and that females living in groups have larger home ranges than they would need if they were not sharing it with others. We also know that frugivorous species typically have larger home ranges than folivorous species (Clutton-Brock and Harvey 1977a), which suggests that the type of food matters, too.

These facts all tell us that individuals are responsive to the foods in their home ranges; however, while this information has been useful in directing us to investigate in finer detail questions about group size, diet, and habitat quality, they do not explain settlement patterns, extent of home range overlap, why female primates live alone or with others, or why different species have certain ranges of group sizes. After all, frugivores are spread across all types of social organizations. These are key questions that have been asked since the beginning of socioecology as a discipline, and they still have not been adequately answered, largely because I think we have been barking up the wrong tree, so to speak, with a focus on predation and less relevant aspects of food and behavior.

To start with, although females are considered the "ecological sex" (Wrangham 1980; Gaulin and Sailer 1985; Gordon et al. 2013), we need a more useful basic premise from that insight than the common phrase, *female reproductive success is limited most by food*. What specifically about food limits their reproductive success? Is it food distribution, food abundance, nutrient quality, rare nutrients, nutrient mixing, fallback foods, or something else? I think these are too variable over time, some even in one home range, let alone an entire species, to determine social organizations, which are highly consistent over time. More consequential may be how much land females need and how they use the land to support their reproductive success. So, let us start with a more basic premise, that *female reproductive success is limited most by the amount of land available to them*. This circumvents questions about what it is about food that limits female reproductive success and will lead us to a richer understanding of the relationship between females and ecology than the finding that home range size is positively associated with body mass or group size (McNab 1963; Milton and May 1976; Clutton-Brock and Harvey 1977a).

It is important at this point to clarify what I mean by "reproductive success." Fieldworkers often measure it by the number of offspring an individual has had over time, which, practically speaking, means the duration of the field study. With funding in short supply and considering the slow maturation rates and long lifespans of most primates, that is the best most of us can do. Measuring reproductive success that way is really just measuring reproduction however, which simply means having offspring. If those offspring die before they can reproduce, the mother's genes will go nowhere. It will be as if she had not reproduced at all. Actual reproductive *success* occurs when those offspring also grow up and reproduce. This distinction is critical to understanding social organizations from the perspective of space use because it means a female may need a larger area of land than what she herself needs to reproduce. In other words, she may need to expand her home range to accommodate the reproductive needs of her mature offspring (Waser and Jones 1983). Consider home range expansion under these conditions as a form of parental investment (Trivers 1972).

As the epigraph at the beginning of the chapter suggests, examining whether an adult, reproductive female shares her entire home range— or part of it or none of it—with other reproductive females (often her daughters) is the first step toward being able to explain the diversity of primate social organizations. With that in mind, I introduce the Variable Home Range Sharing model in this chapter, in which we examine the ways that adult female primates partition their space and how this information can be used to create a new classification system for primate social organizations that is based on functional explanations rather than on simple descriptions of the numbers of adult males and females sharing their space together.

Home Ranges: Minimally, Somewhat, or Entirely Shared?

Focusing on the female home range as the unit of interest, maps from studies over the years show that there is wide variation in the extent to which adult females share their home ranges with other females. Broadly speaking, however, female home ranges may be shared (1) minimally, if at all; (2) somewhat; or (3) completely with other adult females. In the latter case, females sharing the same home range typically live together in groups. In most species, females largely determine their home range boundaries—and thus the extent of home range sharing—while in a few, males control the home range boundaries via patrolling and aggression (Williams et al. 2004; Lemoine et al. 2020).

Under the Variable Home Range Sharing model, females can be classified into seven novel categories that reflect variation in female land use based on whether they determine their home range boundaries, how much of the home range is shared, and the dispersal characteristics of females. Please note that their active period during the diel cycle is not important at this point but will be later when we discuss solitary foraging versus group living. Figure 3.1 provides a dichotomous key to facilitate the placement of females into these categories with only this information at hand by providing a series of increasingly exclusive binary choices. To use the key, begin by deciding if the species in question fits one or the other alternative in step 1 and continue to work down until the type of female social organization is revealed. Those well-versed in primate socioecology will find some traditionally important characteristics either not present or at unexpected placements in the key. For instance, whether or not adult females who share their home range do so with female relatives is irrelevant in identifying female social organizations until step 6. After your species is revealed as having a particular type of female social organization according to the dichotomous key, check it against the expanded descriptions of female social organizations that follow to confirm its assignment.

One important caveat: *my proposed names for the various categories are not intended to imply psychological motivations or personality traits.* I chose the terms to facilitate a more intuitive understanding of variation in the extent of home range sharing and, by extension, the female's ability to expand her home range to include more area than what she herself needs for her own reproduction. One of the benefits of this classification system is that the needed information is now available for many species.

Dichotomous key for classifying female social organizations under the Variable Home Range Sharing model

1a. Adult female home range size expanded by males via boundary patrolling and aggression, and aggression toward females that restricts their movements.

Examples: chimpanzees, spider monkeys

1b. Adult female home range size not determined by males in the manner described above – **go to step 2**

2a. Adult females do not share their home range with other adult females.

Examples: fork-marked lemurs, red titi monkeys

2b. Adult females share their home range with other adult females – **go to step 3**

3a. Adult females share some but not all of their home range with at least one other adult female.

Examples: gray mouse lemurs, orangutans

3b. Adult females share their entire home range with other adult females – **go to step 4**

4a. Only one breeding female per home range (other adult females are non-breeding).

Examples: golden lion tamarins, common marmosets

4b. More than one breeding female per home range – **go to step 5**

5a. Female natal dispersal associated with aggression from other females who share the same home range.

Examples: Milne-Edward's sifakas, red howler monkeys

5b. Female natal dispersal not associated with aggression from other females who share the same home range – **go to step 6**

6a. Adult females typically share their entire home range with female relatives.

Examples: olive baboons, patas monkeys

6b. Adult females typically share their entire home range with female non-relatives.

Examples: mountain gorillas, red colobus

Figure 3.1 Dichotomous key using behavior to classify social organizations of female primates.

Categories of Female Social Organization

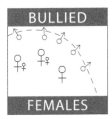

Bullied females are those whose home range boundaries are expanded not by their own volitional behavior but by male behavior via boundary patrolling and aggression, and aggression toward females that restricts their movements. No traditional socioecological classification exists for these females, but some have been recognized as living in fission-fusion societies, or communities. East African chimpanzees may be a good example. In chimpanzees, the community's home range is determined by males, not females (Williams et al. 2002; Mitani et al. 2010). In addition to adjusting their space use relative to food resources and other females (Emery Thompson et al. 2007; Murray et al. 2007; Pusey and Schroepfer-Walker 2013), females adjust their space use in response to male–male competition over community boundaries (Williams et al. 2002; Mitani et al. 2010). The same may apply to Central American and white-bellied spider monkeys, *Ateles geoffroyi* and *A. belzebuth*, respectively (Aureli et al. 2006; Shimooka 2005; Wallace 2008; Ramos-Fernández et al. 2013). Female adjustments of space use in response to male aggression also occur in bottlenose dolphins, *Tursiops* spp. (Wallen et al. 2016), which have often been viewed as convergent with chimpanzees in their social organization and behavior (Pearson 2011).

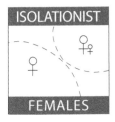

Isolationists do not share their home range with other adult females. In traditional socioecological terms, these are represented by pair-living primates and some, but not all, solitary foragers. Isolationists can be nocturnal or diurnal, and they occur in all major primate radiations. Examples of Isolationists include pale fork-marked lemurs (Schülke and Kappeler 2003; Schülke 2005); Mysore slender lorises, *Loris lydekkerianus* (Nekaris 2003); slow lorises, *Nycticebus coucang* (Wiens and Zitzmann 2003); fat-tailed dwarf lemurs, *Cheirogaleus medius* (Müller 1998); giant northern mouse lemurs, *Mirza zaza* (Rode et al. 2013; Rode-Margono et al. 2016); red-bellied lemurs, *Eulemur rubriventer* (Tecot et al. 2016); indris, *Indri indri* (Bonadonna et al. 2017, 2020); spectral tarsiers, *Tarsius spectrum* (Gursky 2010); red titi monkeys, *Plecturocebus discolor* (Van Belle et al. 2016, 2020); and siamangs, *Symphalangus syndactylus* (Lappan et al. 2021). Isolationists are typically unable to expand their home range enough to accommodate other adult females. Territorial behavior, coupled with living in habitats saturated with other home ranges, appear to make it difficult for females to expand their home range enough to include other adult females (Mitani 1990; Bonadonna et al. 2017, 2020; Lappan et al. 2021), but I will argue in the next chapter that the ways in which females move in their habitats more directly determine the extent to which home range expansion is possible.

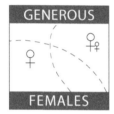 Generous females share some, but not all, of their home range with at least one other reproductive female. Because there is some home range overlap, home ranges are expected to be larger than what an individual reproductive female needs for herself, suggesting that females can expand their home range to some extent to accommodate additional reproductive females. Indeed, home range size for the gray mouse lemur, a typical Generous female, increases as more female home ranges overlap (Dammhahn and Kappeler 2009), and female Allen's galagos, *Sciurocheirus* (formerly *Galago*) *alleni*, expand their home range when they share it with another female and then reduce it when the female leaves (Charles-Dominique 1977). Home ranges of Generous females are often partially shared with adult daughters (Charles-Dominique 1977; Radespiel et al. 2001; Ashbury et al. 2020). Generous females may interact in areas of overlap, even traveling together at times. This behavior has led some to describe Sumatran orangutans, *Pongo abelii*, as living in a fission-fusion society (van Schaik 1999; Singleton and van Schaik 2002). In traditional socioecological terms, however, Generous females, including Sumatran orangutans, are solitary foragers and, with the exception of orangutans, are not active during the day. Besides gray mouse lemurs (Radespiel 2000; Radespiel et al. 2001;

Dammhahn and Kappeler 2009; Wartmann et al. 2010) and Allen's galagos (Charles-Dominique 1977), other examples include Bornean orangutans, *Pongo pygmaeus* (Rodman 1973; Singleton and van Schaik 2002; Knott et al. 2008; van Noordwijk et al. 2012; Ashbury et al. 2020), and Zanzibar galagos, *Galagoides zanzibaricus* (formerly *Galago zanzibaricus*) (Nash and Harcourt 1986).

Intolerant females are similar to Isolationists in that their home range does not include other breeding females. However, Intolerant females will share their entire home range with other adult *non-breeding* females. Additional females who share the same home range may undergo reproductive suppression or, if they do manage to breed, may lose their offspring to infanticide by the higher-ranking reproductive female in the shared home range. In rare cases when home ranges of Intolerant females do support two successfully reproducing females, those home ranges are expected to be larger than home ranges with one reproductive female, as was reported for golden lion tamarins, *Leontopithecus rosalia* (Hankerson and Dietz 2014). In traditional socioecological terms, Intolerant females often live in extended family groups and are represented by marmosets and tamarins (the callitrichines).

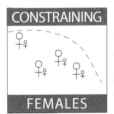

Unlike Isolationists and Intolerant females, Constraining females share their entire home range with other reproductive females. Thus, home ranges are expected to be larger than what an individual reproductive female needs for herself. Home range size is often positively correlated with the number of individuals sharing the home range (Sekulic 1982b; Mertl-Millhollen 1988; Teichroeb and Sicotte 2009; Agostini et al. 2010). However, the extent of home range expansion and sharing is not unlimited. It appears to be constrained based on the observation that targeted aggression toward some females by other females in the group is more likely to occur as the number of reproductive females or total group size increases, resulting in the dispersal of some females from the home range. Constraining females live in small, cohesive, multi-female groups. (The causes of group living will be discussed in chapter 7.) Constraining females are exemplified by Milne-Edward's sifakas (Wright 1995); red-fronted lemurs, *Eulemur rufifrons* (formerly *E. rufus*) (Kappeler and Fichtel 2012; Prox et al. 2023); red howler monkeys (Crockett 1984; Crockett and Pope 1993; Pope 2000); ursine colobus, *Colobus vellerosus* (Teichroeb et al. 2009); and western black-and-white colobus, *Colobus polykomos* (Korstjens et al. 2005, 2007).

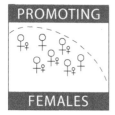

Like Constraining females, Promoting females share their entire home range with other reproductive females. In addition, home ranges of Promoting females typically expand with greater numbers of females in them (DeVore and Hall 1965; Struhsaker 1967; Takasaki 1981; Johnson 1985; Chism and Rowell 1988; Butynski 1990; Wieczkowski 2005). They seem to be less restricted than Constraining females in their ability to expand their home range however, because targeted aggression with female eviction is rare among Promoting females. While minimum female group size does not always differentiate Constraining from Promoting females for a variety of proximate reasons, e.g., demographic changes, predation, or group fissioning (Nash 1976; Isbell et al. 1991; Janmaat and Chancellor 2010), less restricted home range expansion allows for larger maximum group sizes compared to Constraining females, with home range size positively correlated with group size. As with Constraining females, Promoting females are traditionally classified as living in cohesive multi-female groups (Wrangham 1980), but they may also live in multi-level societies. Promoting females are exemplified by chacma baboons (Davidge 1978; Marais et al. 2006; Hoffman and O'Riain 2012), geladas, *Theropithecus gelada* (Dunbar and Dunbar 1974), patas monkeys (Chism and Rowell 1988), and other cercopithecine primates.

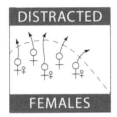

Home ranges of Distracted females are shared entirely with multiple other reproductive females, but females are not bound to a particular home range and may move between home ranges over time. The Distracted female category allows for the possibility that, in some species or even in different populations of the same species, other factors such as predation, infanticide, or the risk of inbreeding depression may limit reproductive success more than access to land and the food on it. Distracted females are distinguished from other types of females in their space use and behavior by directing their efforts to minimize the costs of these other factors. Distracted females live in multi-female groups and have been called non-female-bonded (Wrangham 1980) or Dispersal-egalitarian (Sterck et al. 1997). Mountain gorillas and some populations of ashy red colobus monkeys are examples of Distracted females.

Constraints on Home Range Expansion

While home range *size* is malleable within and across populations, the extent of home range *sharing* seems to be consistent at broad levels within and across populations of the same species, and even across species within a given genus. Thus, for example, under stable conditions, home ranges of slender lorises, indris, and titi monkeys have not been observed to accommodate more than one adult female permanently wherever those primates live (Nekaris 2003; Bonadonna et al. 2020; Van Belle et al. 2016, 2020). Similarly, under stable conditions, home ranges of gray-cheeked mangabeys and ring-tailed lemurs, *Lemur catta*, have always been observed to accommodate more than one adult female (Waser 1977; Sauther et al. 1999).

Of course, exceptions are bound to occur because nature is not perfect. For example, golden lion tamarins are considered Intolerant because home ranges can include a single reproductive female along with non-reproducing adult females, but sometimes two reproductively active females share the same home range. As expected, however, home ranges with two such females are larger than those in which only one reproduces (Hankerson and Dietz 2014). Similarly, female equatorial saki monkeys, *Pithecia aequatorialis,* typically do not share their home ranges with other females (Van Belle et al. 2018) and so are considered Isolationists; however, in a rare case, one adult female twice shared her home range with another adult female, each one presumably her daughter. Each time, both adults reproduced (Van Belle et al. 2016), and in at least one of those years (the analysis presented was unclear for the other year), their home range expanded from the previous year (Van Belle et al. 2016, 2018). Female gibbons have long been considered pair-living, and here they would be considered Isolationists because their home ranges are normally limited to one adult female; however, more recent studies have revealed that, in black-crested gibbons, *Nomascus* (formerly *Hylobates*) *concolor,* and other congeners, home ranges are often fully shared by two adult females (Guan et al. 2018). In one long-term study, membership of adult females sharing the same home range was stable, and both females reproduced (Fan and Jiang 2010, Hu et al. 2018). Perhaps females in this genus of gibbons are better considered Constraining than Isolationists as adult, reproducing females are also

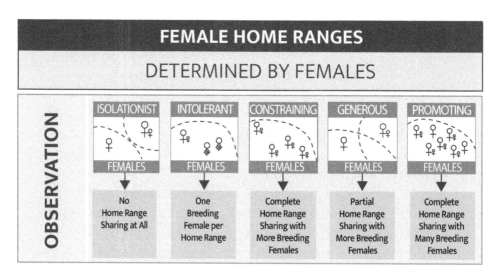

Figure 3.2 Observations of the extent of home range sharing with other adult, reproductive females can be used to functionally categorize the social organizations of female primates that control their home range boundaries.

mother–daughter pairs (Hu et al. 2018). In any case, home ranges in black-crested gibbons with two adult females are, like the exceptions in Isolationist and Intolerant species, larger than those with one adult female (Guan et al. 2018).

In short, we should not be surprised to find exceptions when we study animals long enough, but exceptions such as these do not doom the Variable Home Range Sharing model. Indeed, the exceptions provide supporting evidence of the value of land to female reproductive success because they confirm that home range size is sensitive to the number of reproductively active females present in the home range. This is the case even in species such as golden lion tamarins, equatorial sakis, and black-crested gibbons that can be viewed as territorial (Peres 1989; Van Belle et al. 2018; Hu et al. 2018) and limited in their home range expansion because of their neighbors' contiguous home ranges (Brockelman et al. 1998; Bonadonna et al. 2020).

Main Points

Home ranges are made up of movements of individuals over time. How and when individuals move about their home ranges determine the frequency of interactions they have with conspecifics, resulting in the emergent property we call "social organization." The Variable Home Range Sharing model proposes that female social organizations can be classified into seven types based largely on the degree to which females are able to control and share their home range with other reproductive females.

The extent of home range sharing is important because it indicates the ability of females to expand their home ranges to accommodate those females. This classification system improves upon the social classification system currently in place because it is functional and based in ecology (Figure 3.2), whereas the social classification system is purely descriptive and focuses on the number of adult males and females sharing their space.

CHAPTER 4

Movement Strategies Help Identify Constraints on Home Range Expansion and Their Importance for Female Social Organizations

> The behaviour of females must be an adaptation that maximises their foraging efficiency and their success at raising offspring.
>
> —P. S. Rodman (1973:206)

The previous chapter emphasized the importance of the extent of home range sharing in classifying female social organizations. Because home ranges are made up of a series of daily (or nightly) movements, we will examine in this chapter how females move through that space and how their movement strategies might affect their ability to expand their home ranges to accommodate other adult, reproductive females.

If females can expand their home range beyond what they themselves need to satisfy their own needs for reproduction, then they should be able to share some or all of their home range with other reproductive females. Evolutionary theory expects those other females to be adult daughters, ideally. Longitudinal studies and genetic analyses confirm that, when adult females share some or all of their home range with at least one other reproductive female, they are often mothers sharing with adult daughters (Radespiel 2000; Radespiel et al. 2001; Löttker et al. 2004; van Noordwijk et al. 2012; Ashbury et al. 2020). By continuing to use at least part of the natal home range, adult daughters have a better chance of avoiding predators, they already know where many of their stationary foods are located, and they already know their neighbors (Isbell et al. 1991; Isbell and Van Vuren 1996). All of these factors are expected to improve their own chances of reproduction and thus the reproductive success of their mothers, as benefits of locational philopatry (Waser and Jones 1983).

Given the potential benefits of philopatry, the risks of dispersal, and the dependence of the mother's reproductive success on her offspring's success in reproducing, one would think selection would favor mothers who encourage their daughters to stay and reproduce entirely in the natal home range. All it requires is for the mother to expand her home range. For some females though, it may not be that simple. The existence of so many species in which home ranges include only one adult female suggests that expanding the home range may be more difficult than it seems, at least for them. Female offspring dispersal, adult offspring philopatry but without reproduction, or offspring philopatry but poorer maternal reproduction may all be signs of an inability of mothers to expand their home ranges sufficiently to accommodate their adult daughters' reproduction while also maintaining their own future reproduction. Female pale fork-marked lemurs are excellent examples of this. They are Isolationists with home ranges that are restricted to one adult female, but offspring have been documented to delay their dispersal and remain in the natal home range as adults. In those cases, the mother's reproduction declined or stopped altogether (Schülke 2003). As expected of Isolationist females in general (because they cannot expand the home range easily), home range size was not correlated with the number of individuals sharing the home range (Schülke 2003). The question is, what keeps fork-marked lemurs and other Isolationists from being able to expand their home ranges?

The traditional explanation is that, in populations where home ranges are bordered by other home ranges that are successfully defended, there may be little room or opportunity to expand (Brockelman et al. 1998; Bonadonna et al. 2020). In such cases, we can expect to see home ranges abutting one another. I believe, however, that there is a stronger ecological constraint underlying this social constraint, which can be detected when there are gaps into which home ranges could expand, and yet do not. Normally, siamang home ranges abut, but in one exceptional case, a home range remained stable for many years despite having the room to expand (Palombit 1994). In slow lorises, fork-marked lemurs, and white-footed sportive lemurs, *Lepilemur leucopus*, male home ranges are larger than those of adult females, and yet they do not usually overlap with more than one adult female (Wiens and Zitzmann 2003; Schülke and Kappeler 2003; Drösher and Kappeler 2013). This leaves "empty" areas within adult male home ranges into which females could expand their home ranges, and yet they do not. I propose that the ecological constraint lies in how females move in their environment.

Movement Strategies in Relation to Home Range Expansion

As discussed in the previous chapter, locational dispersal, i.e., movement out of the natal home range and into unfamiliar areas, is fraught with risks, including an unfamiliarity with the locations of foods, which can lead to reduced foraging efficiency that could ultimately affect reproductive success (Ali 1981; Pope 1989; Isbell and Van Vuren 1996; Owen-Smith 2003; Schülke 2003; Pinter-Wollman et al. 2009; Clutton-Brock and Lukas 2012). If we acknowledge reduced foraging efficiency as a risk of locational dispersal, then it means we also expect animals know where their foods are when they are in familiar areas and, consequently, that they are more efficient in getting their food in those areas.

Home range expansion by female owners of those home ranges is similar to dispersal in that it requires the females to move into unfamiliar areas where they do not know the locations of their foods. In those new areas, females are likely to forage less efficiently until they spend more time there (Isbell and Van Vuren 1996; Scarry et al. 2023). Females will not all equally suffer reductions in foraging efficiency in new areas, however. Although no females can know the locations of their foods ahead of time in unfamiliar areas, foraging in unfamiliar areas will be less efficient for those who are more dependent on prior knowledge of the locations of their foods (Schülke 2003; Isbell 2004). The more they rely on prior knowledge to obtain their foods, the more difficulty they will have expanding their home range and sharing it with other reproductive females without a cost to their own reproduction.

I suggest that we can use movement strategies to distinguish females who are dependent on prior knowledge of their food from those who are not (as much). Simply put, females who depend on prior knowledge of the locations of their foods are expected to move directly to those locations. In contrast, females who do not require knowledge of the locations of their foods ahead of time can opportunistically search for (and often fail to find) food as they move along. Since search and failure are part of their foraging repertoire even in familiar areas, such females are predicted to suffer smaller reductions in foraging efficiency in new areas than those who need to know where their foods are located. Females who are not dependent on prior knowledge of the locations of their foods are therefore expected to be able to expand their home ranges and share them with other reproductive females more easily (Figure 4.1).

Animals have morphological adaptations for moving efficiently through their environments. For example, many fish have fusiform bodies for slicing efficiently through the water. Birds have hollow bones, making them lighter and better able to fly (Gleiss et al. 2011). Hominin bipedalism may well have evolved for its locomotor efficiency (Taylor and Rowntree 1973; Rodman and McHenry 1980; Isbell and Young 1996; Sockol et al. 2007). Similarly, we should expect natural selection to have strongly favored in motile animals behavioral adaptations for moving efficiently, particularly to food, because life depends on food and so much time is devoted to getting it.

Movement Strategies in Relation to Categories of Female Social Organization

There are at least three ways that a female can move through her environment that should affect the extent to which she can expand and share some or all of her home range with other reproductive females. The most regimented is *"traplining."* Named to describe the movements of trappers who would check their trap sets repeatedly in the same sequential order, traplining is a movement strategy that works best for animals whose foods, such as nectar and gum, are predictably renewable in the same location (Ohashi and Thomson 2009). Such food sources

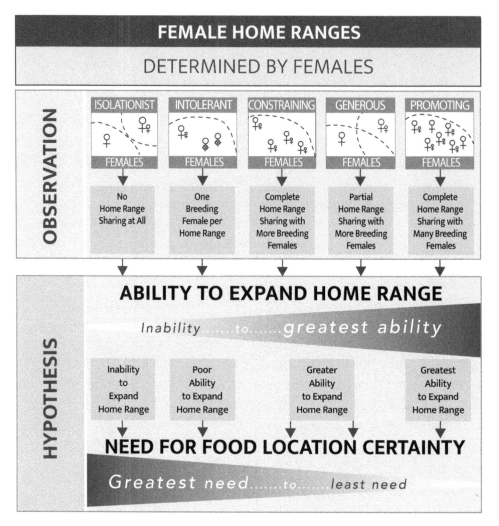

Figure 4.1 The extent of home range sharing is hypothesized to be determined by the relative ability of females to expand their home ranges to share with other reproductive females, which is itself determined by the relative need to know ahead of time where foods are located.

are reliable in space and time, and the most energetically efficient way to take advantage of those properties is to repeatedly return to such foods at appropriate time intervals. Some species of birds (Feinsinger 1976; Gill 1988), bats (Lemke 1984), butterflies (Gilbert 1975), and bees (Heinrich 1976; Ohashi and Thomson 2009) are known to trapline.

Traplining requires the individual to be knowledgeable about predictably renewing resources in space and time. Those that use traplining as their main movement strategy depend on relative certainty in their food locations and so are expected to avoid situations that increase food uncertainty. Two conditions invariably increase food uncertainty. One is moving into new areas, as discussed above. The other is sharing one's home range with individuals whose movements cannot be monitored, because those others could deplete the food if they arrive first. Among female primates, a trapliner is thus expected to benefit from excluding other reproductive females from her home range and minimizing home range overlap with neighboring females. Under these conditions, the home range is expected to be only as large as the female herself needs for her own immediate reproduction. She cannot even share with her adult, reproductive daughters because she cannot expand her home range to accommodate them without reducing her foraging efficiency and the likelihood of producing future offspring.

"Directed travel" differs from traplining only in that movements to foods are not repeatedly sequentially visited. Directed travel occurs when animals travel from one important, e.g., large, food source to another without stopping to feed along the way, and it is most clearly seen when foods are spaced far apart (Chapman 1988; Valero and Byrne 2007; Noser and Byrne 2010). Such movements are often called "goal-directed." Like trapliners, females who employ directed travel as their main movement strategy depend on having confidence in the locations of their foods. They move as if they know the food is ahead of them and in a particular location. They are expected to behave like trapliners in minimizing home range expansion, not allowing other reproductive females to share their home range, and avoiding home range overlap with neighboring females. We might think of traplining and directed travel movement strategies (hereafter simplified to "directed travel" because they produce similar outcomes) as "risk-averse" movement strategies.

Numerous examples exist of primates engaging in directed travel. For instance, fork-marked lemur movements to gum sites have been described as traplining (Schülke 2005). Weddell's saddleback tamarins, *Leontocebus weddelli*, in Bolivia, make non-random, directed, and efficient (i.e., minimizing distance traveled) movements to sequential target trees and sleeping sites (Porter and Garber 2013). Similarly, studies of moustached tamarins, *Saguinus mystax*, and Geoffroy's saddleback tamarins, *Leontocebus nigrifrons* (formerly *Saguinus fuscicollis*), in Peru have demonstrated that they employ directed travel to nectar-producing trees (Garber 1988). White-handed gibbons have been described as regularly traveling long distances directly to preferred food trees (Brockelman 2009; Asensio et al. 2011). Peter Rodman (1979) expressed amazement at observing an adult female orangutan leave a fruiting tree to travel 750 m without stopping only to feed for 20 minutes in another fruiting tree before returning along the same route to the original tree. As a final example, Katie Milton (1980) considered the single-file movements of members in groups of mantled howler monkeys along arboreal routes to their food trees as evidence of "goal-directed travel", which is comparable to the directed travel movement strategy described here.

"*Opportunistic searching*" is the most flexible movement strategy. It involves making short movements while scanning, surveying, or inspecting substrates for food as individuals move along the way to larger food sites. It is often called "foraging" in field studies, but I will not use that term in the context of movement strategies because "foraging" is seldom differentiated from "feeding", i.e., ingestion of foods, with or without forward searching movement (Homewood 1978). Animals that move in an opportunistic searching manner behave as if they do not know the locations of all their foods. Uncertainty and unpredictability are inherent in opportunistic searching, and failure to find food where the individual looks as it moves along is common. For instance, as chacma baboons move along, they sometimes turn over rocks for invertebrates (Davidge 1978; Maré et al. 2021), but they do so independent of what and how much food is under the rocks (Maré et al. 2019, 2021), revealing that they really do not know what they will find and are indeed searching opportunistically. This movement strategy is more compatible with home range expansion and consequent home range sharing with other adult reproductive females than directed travel because moving into new areas necessarily involves not knowing where foods are located.

Regardless of whether they are in familiar or unfamiliar areas, opportunistic searchers will look for and fail to find foods in both, so the inefficiency associated with ignorance of the locations of foods in new areas is likely to be much less than during directed travel. The opportunistic searching movement strategy might be considered "risk-prone" (Figure 4.2).

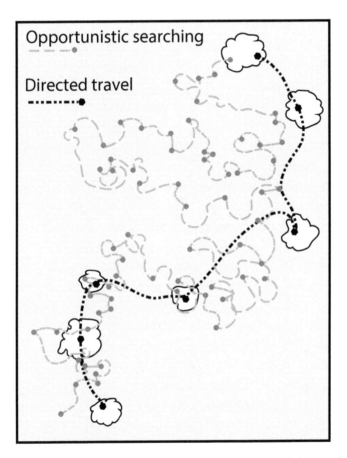

Figure 4.2 Schematic examples of what directed travel and opportunistic searching strategies of primates might look like as primates move through their environments to get to their foods. Directed travel involves movement to food locations without stopping to look for food along the way, whereas opportunistic searching can be more circuitous but, more importantly, involves frequent stops along the way to look for food. The white shapes represent important food trees. The black-filled circles indicate where the directed traveler stopped for food, and the gray-filled circles indicate places where the opportunistic searcher stopped to look for food along its path.

Directed travel has attracted more attention from primatologists because it suggests certain cognitive abilities, whereas opportunistic searching has largely been ignored, perhaps because it seems as if it does not need any special cognitive skills (Noser and Byrne 2007, 2010). Indeed, in their investigation of goal-directed movements and cognition in saddleback tamarins, Leila Porter and Paul Garber (2013) excluded travel of 20 m or less that involved stopping to search for insects, but this indicates that opportunistic searching is also part of the tamarins' repertoire. In studies of chacma baboon cognition, analyses excluded "travel-feeding", defined as when the majority of group members foraged but did not concentrate on the same food source, and with forward progression so slow that they moved less than 20 m in 5 minutes (Noser and Byrne 2007, 2014). Travel-feeding, however, nicely describes the opportunistic searching movement strategy. In their study of chacma baboons, Andrew Whiten and colleagues (1987) specifically avoided comparisons with other baboon species of time spent feeding or moving because it is unclear where "travel-feeding" would fit. From a socioecological perspective, I believe that directed travel and opportunistic movement strategies are equally important to measure.

Most primates probably use both movement strategies but to varying degrees. Right now, we have only qualitative descriptions. During the lean food season, gray mouse lemurs use "efficient travel routes" (Joly and Zimmermann 2011; see also Lührs et al. 2009) to revisit predictable, stationary foods, such as gum sites, and visit them more often than unpredictable, non-stationary foods, such as arthropods (Joly and Zimmermann 2007), indicating a capacity for directed travel. How this compares with their movements during richer seasons has not been examined, but a longer study across seasons reported that they "spent much foraging time searching for prey" (Dammhahn and Kappeler 2008:1582), Arthropods are often unpredictable in space and time (Joly and Zimmermann 2007), and unpredictability encourages opportunistic searching movements. Saddleback tamarins spend about half their time searching for, pursuing, and feeding on arthropods (Garber 1988), suggesting that they also employ opportunistic searching. Siamangs "do not often 'forage'—picking food here and there as they travel between principal food trees" (Chivers 1977:379), suggesting stronger reliance on directed travel. Chacma baboons were described as "both moving and foraging consistently but slowly in small areas of grassland or moving rapidly over some distance" (Marais et al. 2006:75). Orangutan movements have been described as typically following "zigzag routes through the forest feeding at many small food sources . . . before moving off quickly to a new region" (MacKinnon 1974:30). This is consistent with their description as "strongly opportunistic foragers" but with a diet weighted toward fruiting trees (Galdikas 1988:8). Milne-Edward's sifakas make directed movements to their nearest food trees but can also "snack" on leaves as they move between fruiting trees (Erhart and Overdorff 2008a). In my own observations of patas monkeys I found that they spent most of their days wending their way over several hours to a distant resource, such as water, visually searching the area in front of them as they walked and stopping briefly to eat when they found food (Isbell et al. 1998a; see also Hall 1965). These opportunistic searching movements were sometimes interspersed with directed travel when they moved steadily over long distances without searching for food across open ground to reach another grove of trees.

In essence, we do not know the *relative* importance, e.g., frequency, of directed travel and opportunistic searching movement strategies for any primates. This will be important to determine, however, because it will tell us if the Variable Home Range Sharing model is on the right track for explaining primate social organizations. If the relationship between the ability to expand the home range to accommodate additional reproductive females—and therefore, the extent of home range sharing—is dependent on movement strategies, then we should see predictable variation in the extent to which different categories of females engage in each of the two movement strategies. We might predict that, at one extreme, Isolationists and Intolerant females engage most in directed travel because they do not (or cannot) share their home ranges with other (breeding) females. At the other extreme, Promoting females should engage most in opportunistic searching because their home ranges are shared with many other reproductive females, even including relatives beyond their adult daughters. Terrestriality should facilitate

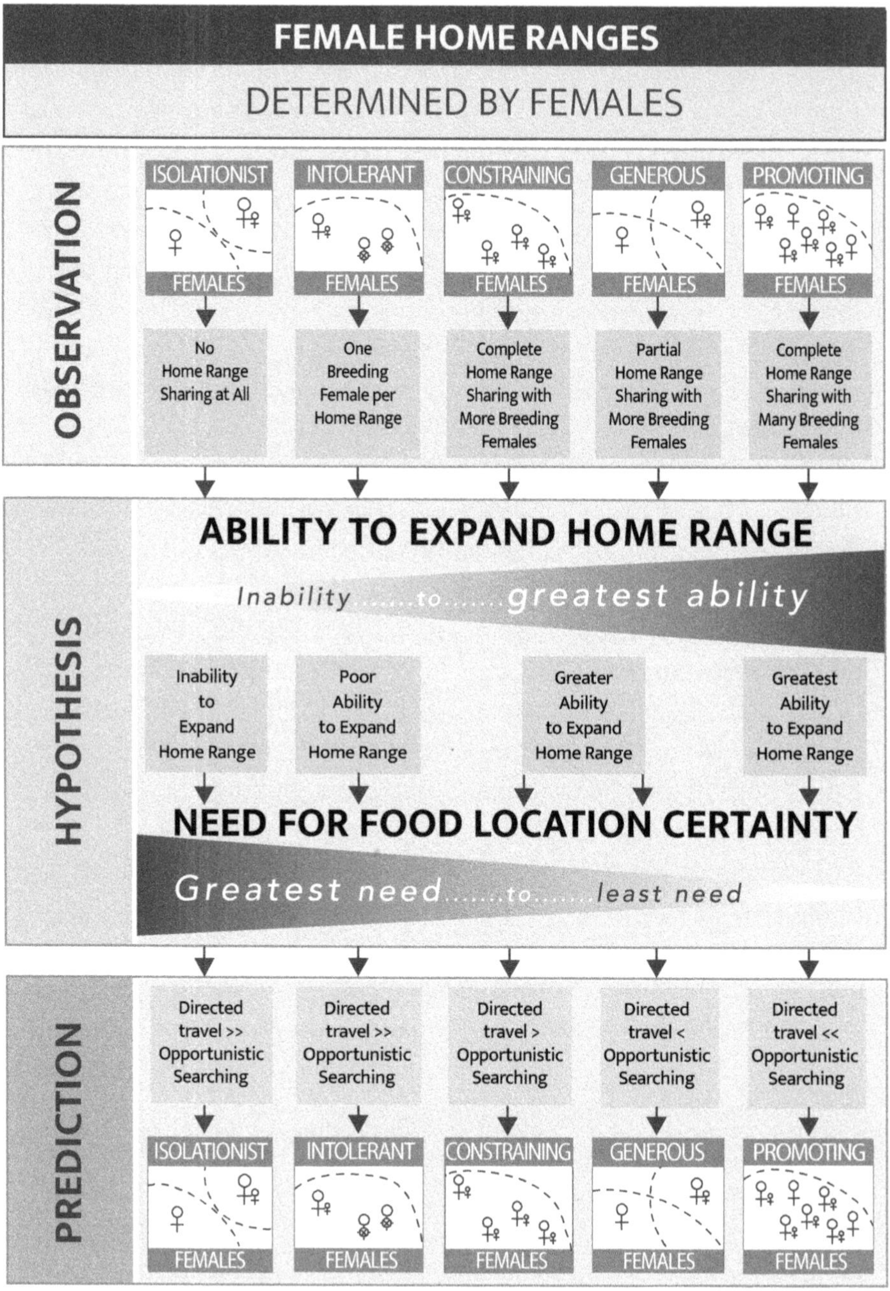

FEMALE HOME RANGES

DETERMINED BY FEMALES

OBSERVATION

ISOLATIONIST	INTOLERANT	CONSTRAINING	GENEROUS	PROMOTING
FEMALES	FEMALES	FEMALES	FEMALES	FEMALES
No Home Range Sharing at All	One Breeding Female per Home Range	Complete Home Range Sharing with More Breeding Females	Partial Home Range Sharing with More Breeding Females	Complete Home Range Sharing with Many Breeding Females

HYPOTHESIS

ABILITY TO EXPAND HOME RANGE

Inability ......to...... greatest ability

Inability to Expand Home Range	Poor Ability to Expand Home Range	Greater Ability to Expand Home Range	Greatest Ability to Expand Home Range

NEED FOR FOOD LOCATION CERTAINTY

Greatest need ......to...... least need

PREDICTION

Directed travel >> Opportunistic Searching	Directed travel >> Opportunistic Searching	Directed travel > Opportunistic Searching	Directed travel < Opportunistic Searching	Directed travel << Opportunistic Searching
ISOLATIONIST	INTOLERANT	CONSTRAINING	GENEROUS	PROMOTING
FEMALES	FEMALES	FEMALES	FEMALES	FEMALES

Figure 4.3 The Variable Home Range Sharing model predicts that the ability to expand the home range to accommodate additional reproductive females, and therefore the extent of home range sharing, is dependent on movement strategies: directed travel is used more by females who are less able to share their home range, and opportunistic searching is used by females who are more able to share.

opportunistic searching, because females moving on the ground are not limited by branch configurations in arboreal environments but are able move in more directions to food than branches allow, giving them more opportunity to encounter foods accidentally. Some Promoting females, e.g., macaques, baboons, and vervets, regularly move on the ground.

In between these two extremes are Constraining females and Generous females. We might predict Constraining females to have more directed travel than Generous females because they appear to limit the number of females in their home range through targeted aggression, as if home range expansion is restricted, and the number of females overlapping the home ranges of Generous females (Singleton and van Schaik 2001; Knott et al. 2008; Dammhahn and Kappeler 2009) may exceed that of Constraining females sharing the same home range. We might also expect that reductions in directed travel are replaced by increases in opportunistic searching in all categories except Bullied and Distracted females (Figure 4.3).

For Bullied and Distracted females, movement strategies cannot be reliably predicted from the extent of home range sharing alone. This is not to say they do not use either movement strategy, just that home range sizes are also affected by conditions either out of their control or unrelated to food (Figure 4.4).

Now that I have set up predictions about the relative importance of different movement strategies according to the new classification system I have presented, the next chapter will take up the challenge of testing those predictions.

Main Points

According to the Variable Home Range Sharing model, the importance of prior knowledge of the location of their foods for females who control their home range boundaries ultimately determines their ability to expand the home range to accommodate other adult females. This is because expanding into new areas invariably entails a process of learning over time where foods can be found, and those who depend on prior knowledge of the locations of foods will struggle more than those who

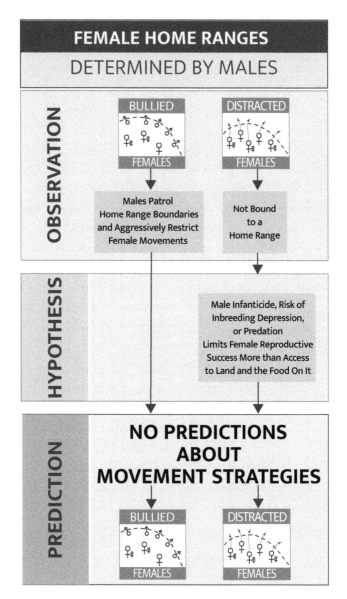

Figure 4.4 Home ranges of Bullied and Distracted females are determined less by food than by non-food-related factors. For Bullied females, males patrol boundaries and act aggressively toward females to limit their movements. For Distracted females, non-food factors such as infanticide, incest, or predation avoidance affect which home ranges females use. No predictions can be made about their movement strategies from the perspective of the Variable Home Range Sharing model.

are not so dependent. Here I hypothesize that the relative importance of prior knowledge can be revealed by how individuals move to their foods: the more often females employ directed travel as a movement strategy, the more important it is for them to have that knowledge. Conversely, the more they employ opportunistic searching as a movement strategy, the less important is prior knowledge of food locations. The consequence is that females who engage in more directed travel will be less able to expand and share their home range with other adult, reproductive females than females who engage in more opportunistic searching, and this has repercussions for female social organizations.

CHAPTER 5

Beginning to Test the Variable Home Range Sharing Model

In other words, chimpanzees traveling between fruit patches typically walk without feeding, whereas bonobos forage as they travel.

—R. W. Wrangham (2000:256)

While my classification system of female social organizations based on the extent of home range sharing with other reproductive females seems straightforward and ecologically sound in theory, the mechanism that I propose to be responsible for that variation (see Figure 4.3) needs further investigation. The challenge now is to develop ways to quantify the relative influence of the different movement strategies so that we can test the predictions of the Variable Home Range Sharing model. Essentially, we need to determine whether there is any correspondence between the extent of home range sharing and the two movement strategies.

Unfortunately, movement strategies have received little attention thus far in primate socioecology. Wrangham (2000) suggested that they might explain the difference in gregariousness of female chimpanzees and female bonobos, but I know of no studies in which the different movement strategies have been quantified. In the future, clever people may come up with direct ways to quantify the relative importance of the different movement strategies, but for now, I offer two indirect quantitative approaches that are worth exploring. These are analyses using random walks and two measures of activity budgets. Of course, right now, with their small sample sizes, they are only a first step that is offered here to point out the possibility that they may be worth building onto with more data in the future.

Indicators of Movement Strategies and Female Social Organizations

Random Walks

Random walks are types of movement behaviors that describe observed frequency distributions of spatial displacements. "Step lengths," which are bouts of near-unidirectional travel of varying distances, document animal movements across the landscape, and statistical distributions fit to these step-length distributions are thought to help reveal the underlying processes (Reynolds 2018). The Lévy walk (or Lévy flight) is a kind of random walk that is characterized by clusters of short step lengths connected by rarer, long step lengths in a power law distribution (Ramos-Fernandez et al. 2004; Reynolds 2018; Reyna-Hurtado et al. 2018), while Brownian movements, another kind of random walk, are characterized by many short step lengths in an exponential distribution (Sims et al. 2014).

Initially, random walks were thought to reflect optimal search strategies when animals move randomly in unfamiliar environments (Viswanathan et al. 1999). In reality, animals with permanent home ranges are not unfamiliar with them. They do not move randomly through their environments, and yet they still express movements that behave like random walks (Schreier and Grove 2010). In fact, recent studies have shown relationships between random walks and the ways in which foods are distributed in the environment (Boyer et al. 2006; Schreier and Grove 2010; Sueur 2011; de Jager et al. 2014). An analysis of step lengths in Central American spider monkeys revealed that they conform to a Lévy walk (Ramos-Fernández et al. 2004). Spider monkey diets are fruit-rich (Chapman 1988), and their "Lévyesque" movements are consistent with movements expected of animals feeding on patchily distributed foods, such as fruiting trees (Boyer et al. 2006), which involve many short step lengths within patches and few long step lengths occurring between patches. In contrast, analyses of hamadryas, *Papio hamadryas*, and chacma baboon movements conform more to a Brownian walk (Schreier and Grove 2010; Sueur 2011). Neither of these species is highly reliant on fruits (Swedell 2002; Henzi et al. 2011), and their Brownian movements are consistent with movements expected of animals feeding on abundant, widely, or more randomly distributed foods (Schreier and Grove 2010; Sueur 2011; de Jager et al. 2014).

Random walks were also examined in a comparative study of frugivorous and folivorous primates (Reyna-Hurtado et al. 2018). Three species of folivorous primates were predicted to move in a Brownian manner because their foods are evenly distributed, while three species of frugivorous primates were predicted to move in a Lévyesque way because their foods are more patchily distributed. Among those classified as folivores, the overall movements of ursine colobus and red colobus were consistent with a Brownian walk, but those of black howler monkeys, *Alouatta pigra*, were not (Reyna-Hurtado et al. 2018). Among those classified as frugivores, the overall movements of spider monkeys and gray-cheeked mangabeys were consistent with a Lévy walk, but those of vervets were not.

In fact, the exceptions above might be better viewed through the lens of the directed travel and opportunistic movement strategies rather than gross dietary categories. Lévy walks appear to me to describe the directed travel movement strategy, and Brownian walks, the opportunistic searching movement strategy (Figure 5.1). "Folivorous" black howlers can also spend substantial time feeding on fruits (41%; Van Belle and Estrada 2020), and they engage in directed travel more when they feed on fruits than on leaves (de Guinea et al. 2021); hence, it might not be surprising that their movements are Lévyesque. Further, since "frugivorous" vervets are often terrestrial, it might not be surprising that their movements are more Brownian because terrestriality facilitates opportunistic searching. Random walks analyses might thus be useful in testing predictions about the relative importance of the two movement strategies for the different categories of female social organizations. Isolationist, Intolerant, and Constraining females are expected to move in a more Lévyesque way, whereas Generous and Promoting females are

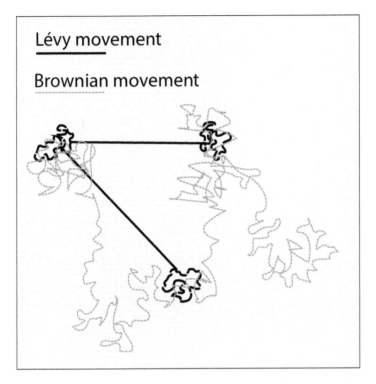

Figure 5.1 Schematic examples of Brownian and Lévy random walks. Compare their paths with opportunistic searching and directed travel paths in Figure 4.2.

expected to move in a more Brownian way. Table 5.1 offers a preliminary examination of random walks as reported in the primate literature, in relation to expectations from the Variable Home Range Sharing model's categories of female social organizations.

As Table 5.1 reveals, the results are mixed. Three of the six species that could be confidently classified moved as predicted by their female social organizations, one was difficult to interpret, and two were opposite that predicted. Among Constraining females, black howlers have Lévyesque (comparable to directed travel) movements, as predicted, but ursine colobus have Brownian (comparable to opportunistic searching) movements (Teichroeb et al. 2009; Dias et al. 2015; Reyna-Hurtado et al. 2018). Among Promoting females, movements of chacma baboons and vervets better approximate a Brownian walk than a Lévy walk

Table 5.1 Species on which analyses of random walks have been conducted, the results of the analyses, and whether the results fit predictions based on the Variable Home Range Sharing classification system of female social organizations.

Scientific name	Common name	Observed random walk	Source	Type of female social organization	Random walk prediction based on type of female social organization	Prediction supported (+) or not supported (−)
Alouatta pigra	Black howler monkey	More Lévyesque	Reyna-Hurtado et al. 2018	Constraining	More Lévyesque	+
Colobus vellerosus	Ursine colobus	More Brownian	Reyna-Hurtado et al. 2018	Constraining	More Lévyesque	−
Semnopithecus entellus	Hanuman langur	Neither Brownian nor Lévyesque[1]	Vandercone et al. 2013	Promoting	More Brownian	−/+
Lophocebus albigena	Gray-cheeked mangabey	More Lévyesque	Reyna-Hurtado et al. 2018	Promoting	More Brownian	−
Chlorocebus pygerythrus	Vervet	More Brownian	Reyna-Hurtado et al. 2018	Promoting	More Brownian	+
Papio ursinus	Chacma baboon	More Brownian	Sueur 2011	Promoting	More Brownian	+
Papio hamadryas	Hamadryas baboon	More Brownian	Schreier and Grove 2010	Promoting/Bullied	More Brownian/No prediction	+/NR[2]
Ateles geoffroyi	Central American spider monkey	More Lévyesque	Ramos-Fernandez et al. 2004; Reyna-Hurtado et al. 2018	Bullied	No prediction	NR
Piliocolobus tephrosceles	Red colobus	More Brownian	Reyna-Hurtado et al. 2018	Distracted	No prediction	NR
Chiropotes sagulatus	Northern bearded saki	More Brownian	Shaffer 2014	Not enough information	?[3]	NA[2]
Semnopithecus vetulus	Purple-faced langur	Neither Brownian nor Lévyesque[1]	Vandercone et al. 2013	Not enough information	?[3]	NA

1. Movement distributions showed a lack of very long movement lengths, which were best described by a truncated power law model rather than purely Lévyesque movements, suggesting that the langurs typically moved to the nearest available food trees. Small home ranges and short daily travel distances may account for the lack of Lévyesque movements (Vandercone et al. 2013), but it is unclear if moving to the nearest available food tree is directed travel or opportunistic searching.

2. **NR**: Outcome not relevant; **NA**: data not available.

3. Data are not available to categorize female social organization.

(Schreier and Grove 2010; Sueur 2011; Reyna-Hurtado et al. 2018), as predicted. Chacma baboons, as well as vervets, spend considerable time on the ground. Promoting Hanuman langurs show a "truncated power law distribution" that is difficult to interpret in terms of directed travel and opportunistic searching movement strategies. The longest movements were relatively short and were attributed to small home ranges and movements to the nearest available food trees (Vandercone et al. 2013). Contrary to expectation, Promoting gray-cheeked mangabeys show a Lévyesque power law distribution (Reyna-Hurtado et al. 2018), but this result may include male movements to foods that females may not prefer (Janmaat and Chancellor 2010) and that take them away from the rest of the group (Olupot and Waser 2005). The movements of female gray-cheeked mangabeys should be examined separately from those of males.

The small sample size and inconsistent results of the association between particular random walks and movement strategies predicted from female social organizations prevent conclusions about whether random walks analyses provide a way to estimate the movement strategies involved in the Variable Home Range Sharing model. That does not mean random walks analyses are not useful for this purpose. I offer these examples to encourage researchers to increase the sample size.

Activity Budgets Based on Time

Field workers frequently collect data on how primates spend their time, and a common way to express this information is with percentages of systematic observations in which the subjects were engaged in a range of activities. These are called activity, or time, budgets. Categories of activities are determined by the researcher and are tailored to what appears to be important to the biology of the animals and the research question. Thus, there is no standard rule for operationally defining categories of activities. Despite this drawback, comparison of certain components of activity budgets offers another potential way to estimate the relative importance of directed travel and opportunistic searching, provided that the categories include directed travel and opportunistic searching. Categories that describe directed travel might be labeled "moving" or "traveling," whereas opportunistic searching might be labeled "foraging." Assuming activity budgets can indeed be used to estimate the relative importance of directed travel and opportunistic searching, Isolationist, Intolerant, and Constraining females are expected to spend more time traveling than foraging, whereas Generous and Promoting females are expected to spend more time foraging than traveling. This is because foraging (opportunistic searching) is most compatible with home range expansion, and home range expansion appears to be easier for Generous and Promoting females.

Fortunately, activity budgets are common in the primatological literature. Using Web of Science and in-text citations from the sources therein, I found 246 records of activity budgets (listed in the Appendix). Additional readings after I finished my literature search indicated that the sample size, while large, is not exhaustive. Unfortunately, most activity budgets could not be used to test the predictions because operational definitions were either not provided or the definitions did not describe directed travel or opportunistic searching movement strategies. For instance, "foraging" was sometimes not distinguished from "feeding" or "moving," and "moving" was sometimes not restricted to "traveling" but also included minor movements as animals fed in a tree. In the end, I was able to use 21 studies because they defined "moving" or "traveling" and "foraging" or "feeding" in ways that appear to describe directed travel and opportunistic searching movement strategies (Table 5.2). Although "foraging" was not separated from "feeding" in five of these cases, the percentage of time spent feeding was equal to or smaller than that of "moving" or "traveling," so it was safe to conclude that foraging (opportunistic searching) would have necessarily involved less time than moving or traveling (directed travel). When possible, I used data from adult females only.

Unless different species within a genus were classified as having different female social organizations, the genus was the unit of analysis since it has been argued that there is a strong phylogenetic footprint in female behavior and social organization (Di Fiore and Rendall 1994; Thierry 2008; Shultz et al. 2011). However, I treated divergent species within a genus as separate data points. This only occurred with *Eulemur*,

Table 5.2 Examples of definitions used in describing directed travel and opportunistic searching.

Definition	Source
Directed travel[1]	
Walks steadily forward without visually scanning the forest floor	Range et al. 2007
Directed movement when not looking for food	Nekaris 2001
Coordinated group progression (often rapid and in single file) or movement between the crowns of trees	Garber 1993
Changing location directionally, in a context of group traveling in a goal-oriented direction	Agostini et al. 2012
Movement with no visual or aural searching for prey apparent	Crompton and Andau 1986
Walk, lope, and run, but not climb or leap (for a terrestrial primate)	Isbell et al. 1998a; this analysis
Continuous, directed movement from one location to another	Rode-Margono et al. 2016
Long uninterrupted bouts of locomotion	McGraw 1998
Travel between trees	Waser 1977
Group moving in a certain direction	van Schaik et al. 1983a
Opportunistic searching[1]	
Moving slowly forward while visually scanning the forest floor, occasionally putting objects in its mouth	Range et al. 2007
Movement associated with looking for food	Nekaris 2001
Localized searches for plant or animal food within the crown of a tree or within a particular microhabitat (e.g., tree trunks, palm crowns, leaf curls)	Garber 1993
Search for food	Crompton 1984; Crompton and Andau 1986
Search, scan, or manipulate food at close range without moving, or scan vegetation while walking	Isbell et al. 1998a; this analysis
Locomotion while feeding; slow quadrupedalism while carefully scanning for fruit and insects; cautious, slow progression	McGraw 1998
Scanning or manipulating the substrate in search of invertebrates	Waser 1977
Group searching for dispersed food items; includes walking, standing, and looking around for food items, as well as handling and eating them	van Schaik et al. 1983a

1. Directed travel was typically labeled as "travel" or "move" in activity budgets, and opportunistic searching was labeled as "forage."

which was represented by two species and two categories of females. *Cercopithecus* was the only other genus represented by more than one species. They are both Promoting females, but their activity budgets were not consistent with each other. Campbell's monkey, *Cercopithecus campbelli*, had strongly divergent scores for traveling and foraging (6.2% and 34.8%, respectively) whereas Diana monkeys, *Cercopithecus diana*, did not (28.5% and 28.3%, respectively) (see Table 5.3). In the end, I decided to exclude Diana monkeys because it is difficult to interpret the biological importance of the very small difference between their traveling and foraging.

I was able to categorize 17 of the 18 genera using the Variable Home Range Sharing model's dichotomous key. I could not categorize *Otolemur* because sources were unclear or conflicting about whether adult females shared parts of their home ranges with other adult, reproductive females (Katsir and Crewe 1980; Clark 1978, 1985; Bearder and Svoboda 2013, 2016). Clark (1978) reported that, even after reaching sexual maturity, daughters may share large portions of the mother's home range and remain closely associated by sleeping and grooming with their mothers but did not mention if these are parous daughters. Bearder and Svoboda (2013, 2016) reported that, while adult females have separate territories that they share with offspring of one or more generations, sleeping sites are not shared with other adult females. Clark (1985) also reported that daughters settle in areas near or overlapping with their mother's home range but that a mother aggressively excluded her daughter. A map shows little home range overlap between two adult females who were aggressive toward each other. Katsir and Crewe (1980) studied the same population as Clark and reported that two adult females did not share their home ranges with others, while two other adult females did, and that females defended home range boundaries. The two females who shared their home ranges nonetheless avoided each other and probably had different core areas. Dispersing males sometimes returned to the study area, but females never did once they left or disappeared, suggesting that female locational and social dispersal may be more common than Clark's studies suggested. Their classification as Isolationist or Generous females awaits further study.

All categories of females were represented among the 17 genera that could be classified. Two genera (*Piliocolobus* and *Procolobus*) were classified as Distracted Females, for which there are no predictions about travel versus foraging time, and so were excluded from further analysis. As mentioned above, *Eulemur* was represented by two categories, and they were treated separately. This left 16 activity budgets for which predictions could be made, of which 13 expressed travel versus foraging percentages in the predicted direction. That is, Isolationist, Intolerant, and Constraining females spent more time traveling than foraging, and Generous and Promoting females spent more time foraging than traveling (Table 5.3). A Fisher's exact probability test revealed statistical support for the predictions (p = 0.04, two-tailed). Thus, there is broad, if not universal, preliminary support for the idea that activity budgets may generate useful quantitative estimates of the relative importance of the two movement strategies across the new categories of female social organizations.

The three genera whose travel versus foraging percentages were not in the predicted direction were *Loris*, *Cephalopachus* (formerly *Tarsius*), and *Lophocebus*. Categorized as Isolationists, *Loris* and *Cephalopachus* were predicted to engage more in travel than foraging, but instead they foraged more than they traveled. It is possible that I misinterpreted the operational definitions of "travel" and "forage." For slender lorises, "foraging" was defined as "movement associated with looking for food," and "travel," as "directed movement when not looking for food" (Nekaris 2001), which may have excluded directed travel to food trees as a later description mentioned travel only in the context of social functions. Travel was "almost always accompanied by scent marking, either urine marking or marking substrates with scent glands. Traveling was more common on the perimeters of the home range and may have a territorial function. Traveling was also used to gain access to opposite sex conspecifics. On several occasions, an animal became active after 2 or 3 h[ours] of remaining relatively motionless. Without any sound audible to the observers, that animal moved in a direct line, often more than 50 m, to an animal of the opposite sex and began a grooming session" (Nekaris 2001:234).

Table 5.3 Species with activity budgets that meet the criteria to test the hypothesis that Isolationist, Intolerant, and Constraining females "travel" (directed travel movement strategy) more than they "forage" (opportunistic movement strategy), and that Generous and Promoting females "forage" more than they "travel."

Scientific name[1]	Common name	Study site	Travel (%)	Forage (%)	Feed (%)	Travel > forage?	Female social organization
Alouatta caraya	Black-and-gold howler monkey	El Pinalito Provincial Park, Argentina	15.0		15.0	Yes	Constraining
Cephalopachus (Tarsius) bancanus	Western tarsier	Sepilok Forest Reserve, Sabah, Malaysia	26.5	60.1	2.1	No	Isolationist
Cercocebus atys	Sooty mangabey	Taï Forest NP,[3] Côte d'Ivoire	10.3	24.5	38.8	No	Promoting
Cercopithecus campbelli	Campbell's monkey	Taï Forest NP, Côte d'Ivoire	6.7	34.8	35.5	No	Promoting
Cercopithecus diana	Diana monkey	Taï Forest NP, Côte d'Ivoire	28.5	28.3	33.2	Yes	Promoting
Chlorocebus pygerythrus (Cercopithecus aethiops)	Vervet monkey	Segera Ranch, Kenya	6.8	7.2	16.6	No	Promoting
Colobus polykomos	Western black-and-white colobus	Taï Forest NP, Côte d'Ivoire	15.1	10.8	34.9	Yes	Constraining
Erythrocebus patas	Patas monkey	Segera Ranch, Kenya	8.6	21.5	15.0	No	Promoting
Eulemur macaco	Black lemur	Ampasikely, Madagascar	32.0, stable across reproductive seasons		15.0, 16.0, 23.0	Yes	Constraining

Alouatta caraya–Eulemur macaco

Prediction supported? (travel versus forage)	Definitions of activities and other relevant information[2]	Sources (first source provides activity budgets)
Yes	Traveling (changing location directionally, in a context of group traveling in a goal-oriented direction), feeding (procuring, handling, ingesting, or chewing any food item), moving (changing spatial position, only including short, nondirectional movements in cases in which the group is engaged in nontraveling activities), resting, social, other. Percentages estimated from graph from maximum of two groups.	Agostini et al. 2012
No	Travel (movement with no visual or aural searching for prey apparent), foraging (active searching for food), feed (not defined), rest, groom.	Crompton and Andau 1986, 1987
Yes	Traveling (long uninterrupted bouts of locomotion), foraging (locomotion while feeding), feeding (not defined), resting. Range et al. (2007) define traveling as "walks steadily forward without visually scanning the forest floor" and foraging (searching) as "moving slowly forward while visually scanning the forest floor, occasionally putting objects in its mouth" and Range et al. (2007) also mention female philopatry. Data from adult females.	McGraw 1998; Range et al. 2007
Yes	Traveling (long uninterrupted bouts of locomotion), foraging (locomotion while feeding), feeding (not defined), resting; data from adult females. Candiotti et al. (2015) mention females are philopatric.	McGraw 1998; Candiotti et al. 2015
No	Traveling (long uninterrupted bouts of locomotion), foraging (locomotion while feeding), feeding (not defined), resting; data from adult females. Candiotti et al. (2015) mention females are philopatric.	McGraw 1998; Candiotti et al. 2015
Yes	Travel (walk, lope, and run, but not climb or leap), forage (search, scan, or manipulate food at chose range without moving or scan vegetation while walking), feed (chew or ingest food without moving or while walking), rest (not moving), other; data from adult females only.	Isbell et al. 1998a, unpublished data
Yes	Traveling (long uninterrupted bouts of locomotion), foraging (locomotion while feeding), feeding (not defined), resting; data from adult females. Korstjens et al. (2005) mention occasional female dispersal. Korstjens et al. (2007) suggest targeted aggression with female dispersal.	McGraw 1998; Korstjens et al. 2005, 2007
Yes	Travel (walk, lope, and run, but not climb or leap), rest (not moving), forage (search, scan, or manipulate food at chose range without moving or scan vegetation while walking), feed (chew or ingest food without moving or while walking), other; data from adult females only.	Isbell et al. 1998a, unpublished data
Yes	Travel (locomotion from one place to another), feed (or forage), rest, affiliative, agonistic, long-distance intergroup interactions, short-distance pacific intergroup interactions, intergroup displays, intergroup chases. Feeding and foraging not defined but percentages indicate travel was greater than either. Percentages by reproductive season. Females can disperse; exclusion of a young female observed; targeted aggression occurs during birth season.	Bayart and Simmen 2005

Table 5.3 *continued*

Scientific name	Common name	Study site	Travel (%)	Forage (%)	Feed (%)	Travel > forage?	Female social organization
Eulemur rubriventer	Red-bellied lemur	Ranomafana NP, Madagascar	22.0		20.0	Yes	Isolationist
Eulemur rufifrons	Red-fronted brown lemur	Ranomafana NP, Madagascar	30.0		20.0	Yes	Constraining
Galago moholi	Southern lesser galago	Transvaal, South Africa	25	60.0	4.0	No	Generous
Leontocebus nigrifrons (*Saguinus fuscicollis*)	Geoffroy's saddleback tamarin	Rio Blanco, Peru	26.0	16.0	15.0	Yes	Intolerant
Lophocebus (*Cercocebus*) *albigena*	Gray-cheeked mangabey	Kibale NP, Uganda	17.2	~10.0	32.5	Yes	Promoting
Loris lydekkerianus	Mysore slender loris	Ayyalur, India	22.3	27.0	0.9	No	Isolationist
Macaca fascicularis	Long-tailed macaque	Ketambe Research Station, Sumatra, Indonesia	18.6	31.1	15.8	No	Promoting

Eulemur rubriventer–Macaca fascicularis

Prediction supported? (travel versus forage)	Definitions of activities and other relevant information	Sources (first source provides activity budgets)
Yes	Travel (including move), feed, rest, other; terms not defined but percentages indicate travel was greater than feeding and foraging. Percentages estimated from graph. Tecot et al. (2016) report minimal home range overlap and that females defend home range boundaries.	Overdorff 1996; Tecot et al. 2016
Yes	Travel (including move), feed, rest, other; terms not defined but percentages indicate travel was greater than feeding and foraging. Percentages estimated from graph. Females observed to disperse through fissionings (Erhart and Overdorff 2008b), but targeted aggression with eviction has been observed among captive animals by Vick and Pereira (1989) and at Kirindy by Ostner and Kappeler (2004) and Kappeler and Fichtel (2012).	Overdorff 1996; Erhart and Overdorff 2008b; Kappeler and Fichtel 2012; Ostner and Kappeler 2004; Vick and Pereira 1989
Yes	Travel (movement without searching), forage (searching for food), feeding (actual consumption of food), rest, groom, play, escape, other; values estimated from a graph. Bearder (1999) mentions daughters may share nest and raise offspring in mother's nest, but Bearder and Martin (1980) report only nulliparous females overlap home ranges with parous females.	Bearder and Doyle 1974; Bearder 1999; Bearder and Martin 1980; Crompton 1984
Yes	Travel (coordinated group progression, often rapid and in single file, or movement between the crowns of trees), foraging (localized searches for plant or animal food within the crown of a tree or within a particular microhabitat, e.g., tree trunks, palm crowns, and leaf curls), feeding (handling or consuming food and included aspects of food preparation such as opening or stripping the husks of large hard fruits), resting. Percentages estimated from graph.	Garber 1993
No	Locomotion (movement between trees), foraging (scanning or manipulating the substrate in search of invertebrates); feeding (eating), movement within trees (4.2%) not counted in other activity categories but likely to be related to food collecting.	Waser 1977
No	Travel (directed movement when not looking for food), forage (movement associated with looking for food), feed (actual consumption of a food item), inactive, groom, other; data from all sexes/ages. Adult female home ranges had minimal overlap.	Nekaris 2001, 2003
Yes	Travel (group moving in a certain direction), search (group searching for dispersed food items; includes walking, standing, and looking around for food items, as well as handling and eating them), feed (group eating from clumped food sources—usually fruit—and includes walking, handling, and eating the food), rest. Values are averaged medians of five groups estimated from graphs.	van Schaik et al. 1983a

Table 5.3 *continued*

Scientific name	Common name	Study site	Travel (%)	Forage (%)	Feed (%)	Travel > forage?	Female social organization
Mirza zaza	Giant northern mouse lemur	Ankarafa Forest, Sahamalaza, Madagascar	65.3			Yes	Isolationist
Otolemur crassicaudatus (*Galago crassicaudatus*)	Thick-tailed greater galago	Transvaal, South Africa	50.0	20.0	5.0	Yes	Isolationist or Generous?
Piliocolobus (*Colobus*) *badius*	Red colobus	Taï Forest NP, Côte d'Ivoire	18.9	15.8	29.1	Yes	Distracted
Procolobus (*Colobus*) *verus*	Olive colobus	Taï Forest NP, Côte d'Ivoire	19.1	12.7	26.5	Yes	Distracted
Saguinus mystax	Moustached tamarin	Rio Blanco, Peru	28.0	16.0	20.0	Yes	Intolerant

1. Scientific names in parentheses are older names used in the cited studies.
2. Definitions of activities in parentheses were taken verbatim from the cited studies.
3. NP: National park.

While Crompton and Andau (1986, 1987) defined "foraging" as "actively searching for food" for Isolationist tarsiers, closer inspection reveals that the active part incorporates a sit-and-wait approach involving scanning rather than searching for food as the animals move. "Tarsiers will cling near ground level on vertical sapling trunks to survey the ground visually and aurally for prey in leaf litter. As Fogden (1974) notes, 10 minutes or more may be spent at a single perch before the animal

Mirza zaza–Saguinus mystax

Prediction supported? (travel versus forage)	Definitions of activities and other relevant information	Sources (first source provides activity budgets)
Yes	Moving (continuous, directed movement from one location to another), feeding and foraging (movement associated with searching for food or actual consumption of food items), resting, grooming, social. Combined, foraging and feeding contributed much less (14.6%) than travel (65.3%) to the activity budget. Home ranges of one adult female and a related subadult female overlapped slightly at most.	Rode et al 2013; Rode-Margono et al. 2016
?	Travel (movement without searching), forage (searching for food), feeding (actual consumption of food), rest, groom, play, escape, other; values estimated from a graph. Typically slow movement between gum sites, punctuated with feeding bouts and periods of inactivity.	Crompton 1984
NA	Traveling (long uninterrupted bouts of locomotion), foraging (locomotion while feeding), feeding, resting; data form adult females. Korstjens et al. (2007) mention female dispersal. Korstjens (2001) mentions too few agonistic interactions to create a dominance hierarchy.	McGraw 1998; Korstjens 2001; Korstjens et al. 2007
NA	Traveling (long uninterrupted bouts of locomotion), foraging (locomotion while feeding), feeding, resting; data from adult females. Korstjens and Schippers (2003) and Korstjens et al. (2007) report female dispersal, no female coalitions, and little aggression toward females.	McGraw 1998; Korstjens and Schippers 2003; Korstjens et al. 2007
Yes	Travel (coordinated group progression (often rapid and in single file) or movement between the crowns of trees), foraging (localized searches for plant or animal food within the crown of a tree or within a particular microhabitat, e.g., tree trunks, palm crowns, leaf curls), feeding (handling or consuming food and included aspects of food preparation such as opening or stripping the husks of large hard fruits), resting. Percentages estimated from graph.	Garber 1993; Garber et al. 1984

moves on, but usually the animal will not spend more than a couple of minutes without moving to a new location." (Crompton and Andau 1986). In this case, "foraging" does not actually appear to reflect the opportunistic searching movement strategy.

Finally, gray-cheeked mangabeys spent more time traveling than foraging, contrary to the expectation for Promoting females. Recall that they were determined to move in a more Lévyesque way, which is also not what is predicted for Promoting females. However, both of these results may be because the data analyzed were from the entire group instead of just adult females, an important consideration for gray-cheeked mangabeys. Unusually for cercopithecines, subadult and adult male gray-cheeked mangabeys are only loosely attached to individual social groups (Olupot and Waser 2005; see also chapter 2). Sometimes they move independently of and far away from their group while still staying in their group's home range (Olupot and Waser 2005; Janmaat et al. 2009). Sometimes they feed in fruiting trees as far as 500 m away from the rest of the group (Janmaat et al. 2006). Males are thus not necessarily representative of how females move in this species and including them may skew the results. In a later study of the same population, males indeed spent more time traveling (59%) and less time foraging (39%) than females (29% and 43%, respectively), although these values were restricted to when they were on the ground (Janmaat and Chancellor 2010). Based on these percentages, female gray-cheeked mangabeys do behave as expected of Promoting females, at least while terrestrial. Unfortunately, traveling and foraging were not defined. As an aside, it is worth considering the possibility that sex differences in movement strategies contribute to the loose association between male and female gray-cheeked mangabeys (see also Wrangham (2000) for chimpanzees).

Incidentally, the study of gray-cheeked mangabeys by Janmaat and Chancellor (2010) is also consistent with the hypothesis that expanding into new areas reduces foraging efficiency for directed travelers but not for opportunistic searchers. When a mangabey study group shifted to an unfamiliar area, males (who may engage more in directed travel) appeared to have more difficulty than females (who may engage more in opportunistic searching) in finding individuals of *Ficus sansibarica*, a preferred food tree species. Moreover, foraging efficiency did not decline for females feeding on fruits in the new area. With their expected heavy reliance on opportunistic searching, Promoting females should be buffered more than other types of females from reduced foraging efficiency in new areas.

Many years ago, Katherine Homewood (1978:376) implored us to distinguish foraging ("visual and/or manipulative search of a potentially food-containing substrate") from eating ("the act of ingestion and/or chewing") and to consider feeding as "combined foraging and eating" when we record data on activity budgets. She argued that, because these activities differ in net energy expenditure, they should be considered distinct activities. It is unfortunate that we did not uniformly follow up on her recommendations because they would have been very useful for testing the predictions here. Fortunately, data on activity budgets are relatively easy to collect, and researchers are asked once again to separate foraging from feeding and traveling and to include the activities that describe opportunistic searching movements and directed travel (see definitions in Table 5.2). Finally, it would be helpful to publish activity budgets for adult females separately from those of other age and sex classes. The support in this preliminary analysis for a link between the extent to which adult females share their home ranges with other reproductive females and the extent to which they (or their groups as a whole) engage in different movement strategies is encouraging. It is clearly worth incorporating these changes into future studies to develop a larger sample size and a more robust analysis.

Activity Budgets Based on Distances Moved While Traveling and Feeding

Arguing that energetic concerns are more about distance than about time spent on a given activity, John Fleagle and Russ Mittermeier (1980) offered a different sort of activity budget. They calculated locomotion bout distances during travel and feeding for six sympatric primate species in Surinam. Their "locomotion during travel" appears comparable to the directed travel movement strategy, whereas their "locomotion during feeding" appears comparable to the opportunistic searching movement strategy. By more directly assessing movements, I believe their distance-informed activity budgets catch the essence of movement strategies better than percentage of time spent in foraging and travel and provide more biologically meaningful interspecific comparisons of the relative importance of directed travel and opportunistic searching movement strategies.

Five of their six species could be categorized by type of female social organization, and included Bullied, Distracted, Intolerant, Constraining, and Promoting females. Bullied and Distracted females must be excluded, however, because we cannot use their movements to predict the extent of their home range sharing. For the other types of females, the percentage of locomotion bout distances during travel is predicted to be greatest among Intolerant females, followed by Constraining females, and then Promoting females, as reflected in increasingly greater home range sharing. The results uniformly support the prediction (Table 5.4), but the sample size is exceedingly small. More studies that incorporate their methods are needed to further test the Variable Home Range Sharing model.

Even if a larger sample size were to be supportive, the results from Promoting Guianan brown capuchins, *Sapajus* (formerly *Cebus*) *apella*, warn us that the logic of the Variable Home Range Sharing model may need tweaking. Although the species are consistent *relative to each other* in that Intolerant midas tamarins, *Saguinus midas*, engaged in directed travel the most and Promoting females engaged in it the least, the predicted extent to which directed travel and opportunistic searching are employed *within a given species* may need to be adjusted. Currently, Promoting females are predicted to engage more in opportunistic searching than directed travel, but in this study, Promoting Guianan brown capuchins engaged more in directed travel (60%) than opportunistic searching (40%). The differences in engagement in directed travel versus opportunistic searching may be less drastic than I have proposed for the Variable Home Range Sharing model. Perhaps, as the comparison suggests, all species do more directed travel than opportunistic searching. Activity budgets based on time seem not to support this however, and the two approaches need to be reconciled somehow.

One way to reconcile them might be to recognize that they provide different services. Activity budgets based on distance provide graded measures of the difference between the two movement strategies within

Table 5.4 Bout lengths of locomotion during travel as a percentage of total locomotion during travel and feeding in six primate species, from Fleagle and Mittermeier (1980).

Scientific name	Common name	% Locomotion during travel[1]	Type of female social organization	Prediction based on type of female social organization	Prediction supported (+) or not supported (−)
Saguinus midas	Midas tamarin	76%	Intolerant	% Locomotion during travel greater than other types	+
Alouatta macconnelli (A. seniculus)[2]	Guianan red howler	66%	Constraining	% Locomotion during travel intermediate	+
Sapajus (Cebus) apella	Guianan brown capuchin	60%	Promoting	% Locomotion during travel less than other types	+
Saimiri sciureus	Guianan squirrel monkey	71%	Distracted[3]	No prediction	NA
Ateles paniscus	Red-faced black spider monkey	67%	Bullied[4]	No prediction	NA
Chiropotes sagulatus (C. satanas)	Guianan bearded saki	61%	Not enough information[5]	?	?

1. Percentage of locomotion during feeding = 100% (i.e., total % of locomotion during travel and feeding)–% of locomotion during travel.
2. Scientific names in parentheses are those used in the cited studies.
3. Based on Boinski et al. (2002, 2005) who reported female dispersal without targeted aggression in this species.
4. Based on Shimooka (2005), Aureli et al. (2006), Wallace (2008), and Ramos-Fernández et al. (2013) who reported that home range size in spider monkeys is determined more by males than females.
5. Females share their entire home range with multiple other females (Shaffer 2013), but female dispersal/philopatry has not yet been described.

a species that are also independent of other activities such as resting and grooming. Activity budgets based on time use binary, "more or less" differences between activities to test the predictions, and the movement-based activities are not independent of other activities. Activity budgets based on time may be useful for testing whether or not a particular species is moving as predicted for its particular social organization, while activity budgets based on distance may offer a more standardized approach for making interspecific comparisons. In the species examined here, the Intolerant midas tamarins showed the greatest difference between movement strategies (52% in favor of directed travel), the Constraining red howlers, intermediate (32%), and the Promoting brown capuchins, the smallest difference (20%), as predicted.

Main Points

The Variable Home Range Sharing model is a conceptual model that attempts to describe and explain the diversity of social organizations in primates based on the extent of home range sharing by females. Differences in movement strategies, the proposed underlying mechanism for variation in home range sharing, were examined with analyses of random walks and activity budgets. While the results were generally supportive, they all suffer from small sample sizes. Lévy and Brownian walks are statistical measures of movements that appear to describe directed travel and opportunistic searching, respectively. Activity budgets, measured either in time or distance moved, may also capture the two movement strategies. In terms of time, the percentage of traveling, which indicates the directed travel movement strategy, was greatest among Isolationists, Intolerant females, and Constraining females, as expected. Conversely, percentage of time spent foraging, which indicates the opportunistic searching movement strategy, was greatest among Generous and Promoting females, also as expected. In terms of distance, the difference between percentages of bout distances in locomotion during travel (equivalent to the directed travel movement strategy) and locomotion during feeding (equivalent to the opportunistic searching movement strategy) declined successively from Intolerant to Constraining to Promoting females, as expected. We desperately need to collect the data to conduct more robust tests in the future however, including adding more rigorous operational definitions and separate analyses that include only adult females.

CHAPTER 6

Problems with Predation as a Selective Force on Primate Social Organizations

There is no satisfactory explanation for why some
primates do not live in groups.
—P. M. Kappeler and C. P. van Schaik (2002:712)

Up until now, I have focused on justifying a new approach for catego-
rizing primate social organizations based on the extent of home range
sharing among adult females. Understanding the fundamental cause of
variation in home range sharing is the first step in understanding variation in primate social organizations. I have further argued that paying
attention to how females move through their home ranges is the key to
understanding why some females share their entire home ranges with
other adult females, why some share only portions, and why others share
none at all.

As I described in chapter 3, the extent of home range sharing is expected to depend on how easily females can expand their home ranges
into new areas to accommodate other reproductive females, often their
adult daughters, and that depends on their ability to maintain their own
foraging efficiency in new areas. Recall that I have defined greater foraging efficiency as greater returns in the currency most important to the
animal—such as calories, avoidance of secondary compounds, or mixing nutrients—for the effort invested in the process of acquiring food.

Foraging efficiency is tempered by the relative importance of needing certainty in knowing the location of foods, however. If females require knowing the locations of their foods, they will have a difficult time expanding into new areas while maintaining their foraging efficiency, but those that already incorporate search and failure into their movements will have an easier time of it. I have purposely avoided discussing how home range sharing affects group living—but now is the time because understanding that constitutes the second step in understanding variation in primate social organizations. Here I am defining group living at its most basic: adult females who consistently travel together.

Of course, sharing a home range is a prerequisite for group living—it is impossible for females to consistently travel together when they do not share the same home range. Thus, Isolationists living separately in their own home ranges cannot live in permanent groups with other females. Generous females also cannot form permanent groups because their home ranges are only partially shared, although they may travel together at times in areas of home range overlap. Intolerant, Constraining, and Promoting females can share their entire home range with other females, but there are limits for Intolerant and Constraining females. Intolerant females may be able to expand their home range to accommodate other females, typically their daughters, but only rarely enough to additionally accommodate the offspring of those females. They generally remain as the only breeding female in their home range. Constraining females may be able to expand their home range more to accommodate a limited number of *reproductive* females. Only Promoting females seem able to expand their home ranges enough to accommodate large numbers of other adult, reproductive females. Again, these arguments about home range expansion do not apply to Bullied females, because males control their home range boundaries, or Distracted females, whose occupation of a given home range is determined more by non-food-related factors.

It is only when females can expand their home range to share it entirely with other females that they have the choice of traveling separately or together, at least theoretically. In reality, females sharing the same home range (other than Bullied females) almost invariably travel together. The big question now is what draws them together? In this chapter, I will first summarize the well-entrenched traditional explanation that predation pressure favors both solitary foraging and group living and show from multiple angles why it is unsatisfactory. Then I will propose an alternative explanation that is fully in line with the argument from the previous chapters that foraging efficiency affects females' movement strategies and is ultimately responsible for differences in the extent of their home range sharing and their social organizations.

Predation Pressure: The Traditional Explanation for Both Solitary Foraging and Group Living

The explanation most often invoked and accepted is that predation pressure is responsible for group living, but also for solitary living, with body size and diel activity patterns being important mediators. Specifically, the argument is that weaker predation pressure has favored solitary foraging (in orangutans) or small groups among diurnal primates, while stronger predation pressure has favored solitary foraging among nocturnal primates and large groups among diurnal primates (Crook and Gartlan 1966; Clutton-Brock and Harvey 1977a; Terborgh 1983; van Schaik and van Hooff 1983; Terborgh and Janson 1986; Burnham et al. 2012).

Nocturnal primates tend to be small (although there are also small diurnal primates). According to the predation hypothesis, their small body size puts them at higher risk of predation because a greater variety of predators can eat them, and they employ crypsis as their main anti-predator strategy since they cannot rely on flight or active defense (Clutton-Brock and Harvey 1977a; Terborgh 1983). Crypsis includes being active only at night and being solitary and quiet during the active period to avoid drawing the attention of nocturnal predators, most of which rely on auditory cues to locate their prey. Grouping in small primates would not be favored because they would be noisier as they feed and move about during the active period, thereby attracting predators (Clutton-Brock and Harvey 1977a; van Schaik and van Hooff 1983; Terborgh 1983; Terborgh and Janson 1986; Hill and Dunbar 1998; Kappeler and van Schaik 2002).

For the larger-bodied, diurnal primates, greater visibility makes crypsis a poor strategy, so the alternative is to avoid predators more actively (Clutton-Brock and Harvey 1977a; van Schaik and van Hooff 1983). One such strategy is argued to be group living, because collective participation in vigilance is then possible, resulting in quicker and more reliable predator detection (Alexander 1974; van Schaik et al. 1983b). Vigilance, as an indication of the likelihood of detecting predators, has indeed been found to increase with increasing group size in numerous species of birds and mammals (Caro 2005), including vervets (Isbell and Young 1993a). Larger groups of long-tailed macaques, *Macaca fascicularis*, were found to react from a greater distance than smaller groups to approaching humans, suggesting a group-size effect on vigilance in that population, as well (van Schaik et al. 1983b). In addition, as a matter more of logic or geometry than active behavior, the chances of being the one taken by a predator are reduced by living in groups (the "dilution effect"), and individuals can use others as a shield between themselves and the predator more easily (the "selfish herd effect"; Williams 1966; Hamilton 1971), although the latter explains the formation of temporary aggregations better than cohesive, permanent groups (Williams 1966).

There is no doubt that predation can be a problem for primates. I know this from first-hand experience after having worked at two sites where leopards annihilated the study groups over a short period of time (Isbell 1990; Isbell et al. 2009). In Amboseli, the estimated predation rate in the study population of vervets was 45% in one year alone (Isbell 1990), and at Segera, 48% of the individuals in the larger of two vervet groups died of suspected predation over one six-day period (Jaffe and Isbell 2010). Despite such heavy losses, the evidence of leopard predation was almost always circumstantial because we never saw an attack and rarely found remains. These experiences led me to undertake a study in which the movements of leopards and their primate prey could be monitored simultaneously to gain a better understanding of how this deadly predator can be so effective (see below).

I have also presented a theory that predatory constricting snakes were largely responsible for the origin of primates by affecting their visual sense, and that venomous snakes favored subsequent differentiation of the visual systems of primates on the landmasses of Central and South America versus Africa and Asia (Isbell 2006, 2009). Subsequent research has overwhelmingly supported the primacy of snakes over other animals (including lizards, birds, and cats) as targets of rapid visual detection, at least among catarrhine primates, particularly when visual conditions are challenging (reviewed in Soares et al. 2017; Kawai 2019). One study, for example, found that people shown an array of images that are progressively less scrambled can recognize images of snakes earlier than other objects (Kawai and He 2016). Another series of studies showed that people preferentially detect snakes when images are presented rapidly (Soares and Esteves 2013; Soares et al. 2014) and on the visual periphery (Soares et al. 2014). A highly reliable cue for rapid detection of snakes appears to be their scales. Humans can detect images of snakes and snake scales faster than images of whole lizards and partial skins, and whole birds and partial feathers (van Strien and Isbell 2017), and wild vervets can detect snakes from their scales alone (Isbell and Etting 2017). Underlying this heightened sensitivity to snakes are visually oriented neurons in the subcortical pulvinar nucleus in the brain that respond more strongly and more quickly to images of snakes than to other stimuli (Le et al. 2013) and respond to snakes in striking postures more strongly than to snakes in resting postures (Le et al. 2014). In short, my own research and that of others support the argument that predation is a strong ecological and evolutionary force acting on primates.

All this is to say that I should be the last person to argue against predation as a selective pressure favoring primate social organizations, but here I am. In my own empirical research, I have found support for predation as a selective pressure favoring non-social anti-predator adaptations. In addition to the evolution of visual-system sensitivity to snakes (Isbell 2006, 2009) described above, my studies suggest that locational philopatry (Isbell et al. 1990) and alarm-calling to deter the predator (Isbell and Bidner 2016) provide non-social anti-predator benefits. I do not dispute that there are anti-predator benefits that can be gained socially, e.g., acquisition of appropriate responses to predators by observing others (Seyfarth and Cheney 1980; Cook and Mineka 1989; León et al. 2023). I have not, however, found support for predation as a selective pressure favoring basic socioecologically focused anti-predator adaptations, such as larger group size (Isbell et al. 2018), despite finding greater scanning behavior in larger groups of vervets (Isbell and Young 1993a).

Certainly, large group size did not provide superior anti-predator advantages to the vervets at my Segera study site when the larger vervet group lost 12 (57%) of its 21 members in one month, while the smaller group lost only one (14%) of its seven members (Jaffe and Isbell 2010; also see below). Moreover, my report that crowned eagles spent less time near larger groups of gray-cheeked mangabeys—as determined by the length of time adult males gave alarm calls during focal samples—may have simply been an artifact of sample duration; focal samples of adult males in smaller groups were twice as long as those in larger groups (Arlet and Isbell 2009), so alarm calls could be recorded for up to twice as long in smaller groups. Although primates clearly have numerous anti-predator strategies, including flight, mobbing, and alarm-calling to alert conspecifics, my own research finds no evidence for primate *social organizations* as an anti-predator strategy. As I will attempt to convince you next, neither does the research of others.

But Nocturnal Primates Are Not Cryptic

It is still a common belief that "In animals that are largely solitary, predation risk is unlikely to be a serious concern" (van Schaik 1999:71), but this view needs to be abandoned. Gray mouse lemurs, for example, have among the highest predation rates of all primates, with about 25% of their population estimated to die annually just to barn owls, *Tyto alba* (Goodman et al. 1993). Moreover, as nocturnal primates have become better studied, we have learned that they do not skulk around alone, trying to remain inconspicuous. Most obvious, perhaps, is their frequent vocal communication, including loud calling (Radespiel 2000; Bearder et al. 2002; Gursky and Nekaris 2019), that could attract avian or carnivoran predators. Indeed, galagos are vocal enough that their calls can sometimes be used in the field to differentiate species (Bearder 1999;

Pozzi et al. 2019). When carnivoran and reptilian predators are nearby, galagos do not always remain silent but give alarm calls that attract others to join and mob the predator (Bearder et al. 2002). Alarm calling and mobbing have also been noted for mouse lemurs, pale fork-marked lemurs, and spectral tarsiers (Schülke 2001; Gursky 2002, 2005; Eberle and Kappeler 2008).

In addition to conspicuous calling that could be used by predators to locate them and active displays in the presence of predators, nocturnal primates engage in social activity that could attract predators, as the following examples show. Adult male and female tarsiers spent 28% of their active time within 10 m of each other (Gursky 2002, 2005). Mysore slender lorises spent 38% of their active time within 20 m of another non-infant, and 69% of their social interactions occurred away from their sleeping site, although, as Isolationists, adult females rarely interacted with each other (Nekaris 2006). During the active period for golden brown mouse lemurs, *Microcebus ravelobensis*, social interactions within 5 m of focal animals who were away from the sleeping site occurred at a rate of 1.7 interactions per hour (Weidt et al. 2004). Oliver Schülke (2001) has pointed out already that these are not the behaviors of cryptic prey. Indeed, other than foraging by themselves—which necessarily limits the extent of socializing—and relying more heavily on olfactory communication (Braune et al. 2005; Dammhahn and Kappeler 2005; Drea et al. 2019), nocturnal primates appear not very different from diurnal primates in regularly vocalizing, engaging with predators of appropriate size and level of threat, and interacting with conspecifics when they are nearby. As Claudia Fichtel (2007:613) has succinctly concluded, "solitariness and crypsis can no longer be considered as global, main antipredator strategies for all nocturnal primates." Something else must explain why nocturnal primates are solitary foragers.

Predation and Group Living in Diurnal Primates: Also on Shaky Ground

In attributing group living in diurnal primates to predation pressure, we (1) equate higher predator densities with higher predation risk, (2) assert that smaller-bodied primates and smaller primate groups are more vulnerable to predators than larger-bodied primates and larger primate groups, and (3) expect groups to be larger when predation pressure is greater (Crook and Gartlan 1966; Alexander 1974; van Schaik 1983; van Schaik and van Hooff 1983; Anderson 1986; Terborgh and Janson 1986; Dunbar 1988). These beliefs may seem logical, but many presumed primate adaptations to predation currently have little actual empirical support (Cheney and Wrangham 1987; Isbell 1994; Miller and Treves 2007). The so-called evidence for predation as the force behind group living in diurnal primates comes largely from field experiments using predator proxies—such as humans or recordings of vocalizations played back to primates—and from comparisons with non-primates, not from direct evidence of predator interactions with diurnal primates.

Actual interactions between predators and their primate prey have rarely been observed despite the efforts of many people over many thousands of hours in the field with primates over the years. This does not mean that predation is not important in the lives of primates, it just means that predators are often stealthy, quick, active when observers are asleep, or unhabituated to human presence. Leopards, the predator likely most often responsible for primate deaths from predation in Africa and Asia, are notoriously difficult to catch *in flagrante* for all these reasons (Isbell 1990; Isbell and Young 1993b; Isbell et al. 2018). Let us now turn to problems with the "evidence."

Field Experiments Using Predator Proxies Are Limited in Scope and Can Be Misinterpreted

Greater vigilance is considered a major benefit of grouping and living in larger groups, not just for primates but for many animals, as long as the benefit of vigilance exceeds the greater detectability of larger groups by the predators themselves. Individuals in larger groups also theoretically benefit from greater overall vigilance from the group while reducing their own effort in vigilance, allowing them to engage in other activities, such as feeding (Pulliam 1973; Powell 1974; Hoogland 1979; Barnard 1980; Bertram 1980; de Ruiter 1986; Isbell and Young 1993a).

Humans have sometimes served as a proxy for predators to test the prediction that larger groups detect predators earlier or from a greater distance than smaller groups (van Schaik et al. 1983b; Cowlishaw 1997; Mikula et al. 2018). By remaining unobstructed and walking directly toward groups, researchers then record the distances at which they are detected, as measured by flight initiation. Positive correlations between group size and flight distance are considered evidence for strong predation pressure selecting for increased group size, and the lack of a correlation, evidence for weak predation pressure. In pig-tailed macaques, *Macaca nemestrina*, long-tailed macaques, and vervets, larger groups or subgroups did detect and flee from humans at greater distances than smaller groups (van Schaik et al. 1983b; Mikula et al. 2018). In contrast, in Thomas's langurs and white-handed gibbons, group size had little effect on detection distances (Table 4 in van Schaik et al. 1983b).

The lack of a positive correlation between group size and detection might be expected for gibbons, an exemplar of species that live in small groups presumably in response to low predation pressure (van Schaik and van Hooff 1983). However, their strong responses to visual predator models do not support this interpretation. When shown visual models of tigers, *Panthera tigris*; clouded leopards, *Neofelis nebulosa*; crested serpent eagles, *Spilornis cheela*; and reticulated pythons, *Python malayopython* (formerly *P. reticulatus*), all potential predators of gibbons, they (1) vocalized and increased vigilance in response to all models compared to the control, (2) went lower in the canopy in response to all but the snake model, (3) defecated more frequently in response to the leopard and snake than to other models, (4) dropped more branches in response to the leopard than to other models, and (5) left the area more quickly in response to the tiger model (Clarke et al. 2012). Their reactions do not seem to be those of a species with an evolutionary history of low predation or a poor ability to evolve adaptive responses to it.

A related presumed measure of predator detection is the distance at which animals begin to alarm call at approaching humans. Baboons have long been offered up as exemplars of species that live in large groups in response to high predation pressure from living in open habitats (DeVore 1963; Crook and Gartlan 1966; Dunbar 1988); yet when humans approached unhabituated chacma baboons, larger groups did not begin alarm calling at greater distances than did smaller groups (Cowlishaw 1997).

Although a review of birds and mammals found that larger groups detect predators earlier or from a greater distance, even while individual vigilance declines (Caro 2005), closer examination reveals this group-size effect in mammals only when humans approach in the open or with predators that do not ambush or use crypsis to hunt (see Table 4.3 in Caro 2005). Under obstructed visual conditions and with ambush predators, typical of conditions facing primates, it may be a better strategy to maintain individual vigilance, even in larger groups. Indeed, in a survey of primates, Treves (2000) found that time spent in vigilance did not decrease for individuals living in larger groups except for my study population of vervets living in Amboseli; individual vigilance did indeed decline, while overall vigilance increased in larger groups (Isbell and Young 1993a). That population may be the exception that proves the rule, however—their habitat was extraordinarily open, with few places for ambush predators to hide (Figure 6.1).

Given the difficulty of observing predation itself, our greatest sources of information on predation pressure have come from investigations using visual models for predators and acoustic playback experiments that trick primates into perceiving that a predator is nearby. In addition to white-handed gibbons, many other primates have been found to respond to visual predator models with anti-predator behavior, including tarsiers (Gursky 2006); white-faced capuchins, *Cebus imitator* (formerly *C. capucinus*; Meno et al. 2013); vervets (Isbell and Etting 2017); samango monkeys, *Cercopithecus albogularis* (LaBarge et al. 2021); long-tailed macaques (van Schaik and Mitrasetia 1990); bonnet macaques, *Macaca radiata* (Coss et al. 2007); sooty mangabeys, *Cercocebus atys* (Mielke et al. 2019); and chimpanzees (Crockford et al. 2012).

An iconic example of the use of acoustic stimuli in studies of predation pressure is the research conducted by Robert Seyfarth and colleagues (1980a, b), who showed experimentally that vervet monkey alarm calls function as a warning system to conspecifics—many of whom are relatives—with clear information about the type of predator detected so that group members can respond appropriately. When recorded

Figure 6.1 Amboseli National Park's open habitat provides few places for ambush predators to hide, perhaps favoring a group-size effect on vigilance in detecting predators. The view overlooks the home ranges of several vervet groups at the time their vigilance behavior was studied (Isbell and Young 1993a).

"leopard" alarm calls were played back to the vervets, the most common response was to run into or higher in trees, when "eagle" alarm calls were played back, it was to look up, and when "snake" alarm calls were played back, it was to look down. Since then, numerous other primates, including Verreaux's sifakas, red-fronted lemurs, Geoffroy's saddleback tamarins, moustached tamarins, Diana monkeys, Campbell's monkeys, and bonnet macaques, have been found to respond appropriately not only to conspecific, but also heterospecific, alarm calls (Ramakrishnan and Coss 2000; Zuberbühler 2000, 2001; Fichtel 2004; Kirchhof and Hammerschmidt 2006). Attending to heterospecific alarm calls throws even into question the importance of alarm calls to the evolution of group living. Again, these behaviors merely demonstrate that predation pressure has been important, not that it has been important for the evolution of group living.

One problem with using visual and acoustic models is that they are one-sided, revealing only prey behavior. They simply cannot capture the dynamic nature of interactions between predators and prey, information that is needed to fully understand the selective pressures primates face. Although it has long been known that vervet alarm calls given in the presence of a leopard alerts conspecifics to the threat, as described above (Struhsaker 1967; Seyfarth et al. 1980a, b), it was not known until leopards and vervets were monitored simultaneously that vervet "leopard" alarm calls also function as a predator deterrent. With the aid of camera traps, acoustic recorders, and global positioning system (GPS) collars fitted to vervets and leopards, it became apparent that leopards responded to vervets' alarm calls within minutes by changing direction and moving at least 200 m away (Isbell and Bidner 2016; see Zuberbühler et al. 1999 for the same phenomenon in other primates). Acoustic or visual signals function as a predator deterrent for many non-primate prey species as well (Woodland et al. 1980; Hasson 1991; Caro 1989, 1995, 2005; Clark 2005; Stankowich and Coss 2008).

Even animals that are not social produce alarm calls, and a predator deterrent function might explain how alarm calls initially evolved. Animals that evolved social living then could tack on a social anti-predator function to their alarm calls, as suggested for rodents (Shelley and Blumstein 2005). Incidentally, a predator deterrent function for alarm calls also creates a problem for studies that use the locations of alarm calls to estimate predation risk. The assumption of such studies is that a higher frequency of alarm calls indicates greater predator presence and, thus, greater risk (Willems and Hill 2009; Coleman and Hill 2013; Campos and Fedigan 2014). For alarm calls that do not function as a predator deterrent, this might be a reasonable assumption, although risk might still be lowered if the alarm calls reduce the chance that the predator will be successful. Counter-intuitively, for alarm calls that do serve as a predator deterrent, a higher frequency of alarm calls may actually indicate *lower* risk because the predator gives up and leaves.

Studies relying on proxies for predators of primates are artificial and, as this short review has shown, open to misinterpretation. The full richness of predator–prey relationships can only be revealed when we study primates and their predators at the same time. Only then can we really understand if our assumptions and expectations about the influence of predation on primate social organizations are accurate. This is no easy feat, however; the expense and logistical difficulties alone can be very challenging. Without a strong commitment to clear the roadblocks to carrying out simultaneous predator–prey studies, we cannot truly expect to understand how predation has influenced primate behavior. For now, we must rely on the evidence from just the few simultaneous predator–prey studies that I describe below. They reveal that it is not useful to view "predation pressure" as a unitary phenomenon because predator–prey interactions can be predator-, primate prey-, and habitat-specific. Some of them also challenge with empirical evidence some of our most basic assumptions about the ways in which "predation pressure" has affected primates and their social organizations.

Simultaneous Predator–Prey Studies

Variation in Interactions between Predators and Primates in Different Habitats

Ashy red colobus and western red colobus, *Piliocolobus badius*, were once considered a subspecies of *Colobus badius* (Struhsaker 1975). They are both Distracted females according to the dichotomous key in chapter 3 as both live in large multi-male, multi-female groups with female-biased dispersal (Struhsaker 2010; Korstjens et al. 2022), but they respond differently to chimpanzees, their main predator (Bshary 2007). While we cannot rule out that species differences contribute to this, there is more evidence that habitat differences are responsible (Bshary and Nöe 1997; Bshary 2007). For instance, western red colobus living in Taï National Park, Côte d'Ivoire, a forest with a tall, continuous canopy, move silently away when chimpanzees are farther away, climb higher in the trees if chimpanzees get close enough to be beneath them, and seldom form adult male coalitions to counterattack. For their part, chimpanzees actively search for red colobus groups to hunt, do not discriminate against certain age or sex classes as prey, and often coordinate with each other during the attack (Boesch and Boesch 1989; Boesch 1994). In contrast, ashy red colobus living in Gombe, a forest with a lower, more broken canopy (Boesch 1994), seldom flee from approaching chimpanzees. Instead, adult females climb higher while adult males move lower and frequently form coalitions to counterattack (Stanford 1995). Unlike Taï chimpanzees, Gombe chimpanzees hunt red colobus opportunistically, target immatures, and coordinate less with hunting partners (Stanford et al. 1994). During the two years when I studied ashy red colobus at Kanyawara, in Kibale National Park, I was twice treated as a kind of chimpanzee when coalitions of adult males charged at me from the trees, both times in anthropogenically disturbed areas of the forest where the canopy was broken and low in stature. I escaped by darting into denser vegetation where they could not see me. At Ngogo, also in Kibale, David Watts and John Mitani (2002) confirmed that chimpanzees pay attention to canopy structure, preferring to hunt red colobus in areas with broken canopy and being more successful in those areas. For good reason, red colobus perceive themselves at greater risk from chimpanzees (and humans?) in areas with broken canopy.

Simultaneous study of chimpanzees and red colobus shows us that behavioral interactions between predators and prey are responsive to constraints and affordances in different habitats, even within similar predator–prey systems. However, as the next section shows, sometimes the same predator species behaves in the same way despite great differences in habitat.

Consistency in Interactions between the Same Predator Species and Different Primate Species in Different Habitats

In the same Taï Forest where chimpanzee–red colobus interactions were studied, Klaus Zuberbühler and colleagues (1999) studied leopard interactions with a community of six primate species (two colobus species, three guenon species, and one mangabey species). They radiocollared one leopard and followed her via telemetry at a distance of 30–150 m for 27 days. Visual contact with the leopard and the primates was avoided because none were habituated to human presence. The leopard's behavior, identified as either moving or resting, was inferred from differences in the radio signal, with leopard–primate encounters defined as beginning when the leopard approached and stopped to within 50 m of a primate group and ending when it moved away more than 50 m. Primate groups were identified to species by their vocalizations, with a sudden high rate of alarm calling denoting that they had detected the leopard. Twenty-four encounters occurred, at a rate of 0.89 per day. Of 18 encounters for which durations were determined, the encounters lasted 7–285 minutes. In all of these encounters, monkeys (the species involved was not mentioned) eventually detected the leopard, and once they began calling, in all but 3 of the 18 cases, the leopard left shortly afterward. This was the first evidence suggesting that primate alarm calls function to communicate to the predator that it has been detected. For leopards and other predators that hunt using stealth and concealment, being discovered is surely problematic.

Across the continent from Taï, in a mosaic habitat of semi-arid bushlands and riverine woodlands in Laikipia, Kenya, my colleagues and I captured four leopards and 12 adult female vervets from five groups, fitting them all with collars bearing GPS sensors that allowed their locations to be synchronously recorded every 15 minutes throughout the 24-hour diel period for up to a year (Isbell et al. 2018). At a sleeping site that vervets used on 97% of the study nights, we also set up three camera traps and an acoustic recorder that documented vervet "leopard" alarm calls throughout the night. We found that the more often vervets alarm-called in a given hour, the fewer times leopards were photographed in that hour. Since the camera traps were very near the sleeping trees, this meant that the alarm calls kept leopards away from the sleeping site. Collared leopards corroborated this, as they changed direction and moved away almost immediately after alarm calls began, but continued moving forward in the same direction when vervets did not give alarm calls (Isbell and Bidner 2016). Thus, at least one consistency exists in the leopard–primate relationship across different primate species regardless of the type of habitat and time of day, likely because leopards are ambush predators wherever they live.

Variation in Interactions between the Same Predator Species
and Different Primate Species in the Same Habitat

Despite their generally consistent hunting style, leopards had different relationships with different primate species in the same habitat. In addition to leopards and vervets, we collared six adult female olive baboons from four groups in Laikipia. The processes by which leopards hunt primates there were revealed when collared leopards killed a collared adult female baboon and a collared adult female vervet. In the first case, GPS locations plotted onto Google maps and data from tri-axial accelerometers on the collars showed that a subadult male leopard (named Tatu) detected the baboon (named Thelma) at night on a small kopje (a rocky outcropping) while Tatu was on his way elsewhere. Thelma was likely ill or injured because she had become increasingly separated from her group during the last week of her life, and the kopje on which she settled that night was too small to be a normal sleeping site. After stopping about 275 m from her for an hour, Tatu deviated from his northerly path, and moved east quickly toward Thelma, even crossing a river to get to her. She never moved from her location. After killing Thelma at about 9:00 P.M., Tatu carried her 90 m away to consume her. He stayed in that location for 3.5 hours before moving off (Figure 6.2; see also Figure 6.4), leaving only Thelma's maxilla, some fur, and the GPS collar behind (Isbell et al. 2018).

In the second case, an adult female leopard (named Konda) began moving toward the vervet (named Alan Smith) at midday when they were 234 m apart. Konda may have noticed the vervet group at the Research Center because they were active on top of or near a house that stood higher than the surrounding bushland. Over the next 15 minutes, Konda moved closer until she was 9 m from Alan Smith. Then, 3–5 minutes later, around 12:48 P.M., Konda attacked Alan Smith. Apparently, no vervets saw Konda before the attack because alarm calls began only afterward, as documented by Laura Bidner, a member of my research team, who was fortuitously in a building nearby and heard the alarm calls. Unseen by people at the Research Center, Konda carried Alan Smith more than 235 m away to a lugga (a dry riverbed) (Figure 6.3). Being smaller than an adult female baboon, Alan Smith would

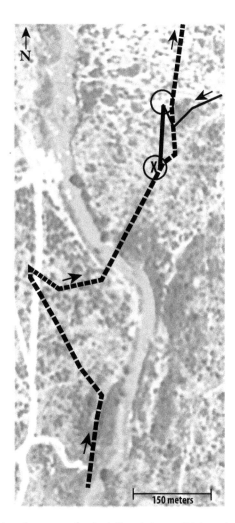

Figure 6.2 The path and process of subadult male leopard Tatu as he hunted and killed female olive baboon Thelma. The heavy dashed line is Tatu's path. The narrow solid line is Thelma's path to the kopje where she slept while she was still alive, and the wider solid line is her path after her death. The open circle shows where Tatu killed Thelma and the circle with "X" marks where her GPS collar was found. The background image is from Google Earth Pro version 7.3.6.9345. The map shows the location during the daytime, but the attack occurred at night.

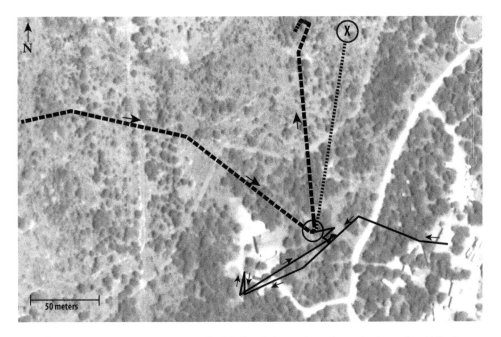

Figure 6.3 The path and process of adult female leopard Konda as she hunted and killed adult female vervet Alan Smith. The heavy dashed line is Konda's path, and the solid line is Alan Smith's path while she was alive, starting at her sleeping site on the roof of a building the night before. The light dashed line is Alan Smith's path after her death as Konda carried her away. The open circle shows where Konda killed Alan Smith and the circle with "X" marks where her GPS collar was found. The appearance of a deviation from Konda's and Alan Smith's paths is likely a result of both artifact in showing straight-line distances between successive GPS fixes rather than actual paths and inherent error in GPS fixes of up to 20 m (Isbell et al. 2018). The rectangular shapes are buildings at the Mpala Research Centre. The background image is from Google Earth Pro version 7.3.6.9345.

have been consumed more quickly, but Konda stayed at that location for 15 hours before finally moving rapidly away, leaving behind only some fur and the collar (Isbell et al. 2018).

The two kills were similar in that both victims were detected from a long distance, and both were carried away from the kill sites. If I had known before that leopards often carry their primate victims far away from their kill sites, my students, field assistants, and I might have found more physical remains at the two other study sites where the primates I studied were subjected to intense leopard predation (Isbell 1990; Isbell et al. 2009; Jaffe and Isbell 2010). An important difference included the timing of the kills: the baboon was killed during the night and the vervet, during the day.

Spatial movements of leopards and the two primate species also revealed that the primates' vulnerability was not the same even though they live in the same habitat with the same individual predators. Importantly, the difference in their vulnerability is not consistent with theoretical assumptions that have been used to formulate the predation hypothesis for group living.

First, leopards preferred to kill baboons rather than vervets even though baboons are more formidable prey because of their willingness to counterattack, their larger body size, and their larger group size, all supposed adaptations to reduce predation. Leopards sometimes traveled far to hunt baboons at night and were more likely to kill baboons than vervets at their most vulnerable time period (night for baboons, day for vervets). While preferring larger prey may make sense from the leopard's perspective, it runs counter to a main assumption of the predation hypothesis that larger primates are less vulnerable to predation than smaller primates (Crook and Gartlan 1966; Clutton-Brock and Harvey 1977a; van Schaik 1983; Dunbar 1988).

Second, greater distance from refuges was not associated with predation events, contrary to the expectation of the predation hypothesis for primates living in more open habitats (Crook and Gartlan 1966; Dunbar 1988; Cowlishaw 1997). Certainly, vervets were safer in the trees than on the ground but only if they were already in them. Closer proximity to trees once vervets were out of them did little to increase their safety because the vervets whose remains confirmed leopard predation died very near trees. Since leopards are ambush hunters, the vervets they killed probably did not have enough time to run into the trees. Rather than safety being a matter of degree, i.e., that greater distance from trees makes vervets more vulnerable, it is more like "all or nothing." Simply being out of trees makes vervets more vulnerable. The irrelevance of greater distance from refuges is even clearer for baboons. They could apparently traverse with impunity thick bush without needing "refuges" during the day but were killed most often at their sleeping site refuges during the night. (Of course, however, there is always the possibility that they would have been at even greater risk had they slept at night on the ground instead of in trees or on cliff faces.)

Third, from brief chance observations over the years, we know that when vervets detect leopards, their typical response is to increase the distance between the leopard and themselves (Cheney and Seyfarth 1981). They do not approach leopards because they are too small to counterattack. Thus, when the synchronous GPS locations over time showed vervets approaching leopards, it meant that they had not yet detected the leopards. This happened in at least 34% of all encounters (operationally defined as proximity within 160 m, the average group spread of the largest group of vervets at the study site). Importantly, larger vervet groups missed detecting leopards as often as smaller groups did, supporting the argument that group size makes little difference in detectability when the predator is an ambush predator (Isbell et al. 2018).

Not only was there no correspondence between group size and the vervets' ability to detect leopards, there was also no correspondence between interspecific group size and the ability to detect leopards. Baboons can chase and even kill leopards (Cowlishaw 1994; Cheney et al. 2004; Isbell et al. 2018); hence, whenever simultaneous GPS locations showed the baboons maintaining their trajectory and passing by a leopard, along with the leopard remaining stationary, it meant that they had not detected the leopard. When the leopard was stationary, it behaved as if it was hiding from the baboons, not as if it had caught an uncollared baboon (because it would have moved away quickly). In contrast, when the baboons curved around the leopard, veered off in another direction, or reversed their direction, it meant that they had detected the leopard. Despite living in larger groups than vervets, baboons missed detecting leopards in 62% of all encounters, almost twice as often as vervets (Isbell et al. 2018). I suspect the baboons had greater difficulty detecting leopards because they often walked on the ground in bushy habitat with limited visibility (Figure 6.4), whereas vervets were more often in trees along the river, giving them a better angle of view of their surroundings (Figure 6.5; note the vervet at the top of the tree). This lack of a correlation between group size and leopard detection,

Figure 6.4 Typical wooded bushland habitat of olive baboons at Mpala, in Laikipia, Kenya. Baboons often failed to detect leopards while walking through this habitat. The small kopje on the ridge (identified by the arrow) is also where a male leopard killed a female olive baboon at night after he crossed the river (in the foreground) to reach her (see also Figure 6.2).

both within and between primate species, does not support the predation hypothesis for group living (van Schaik 1983; van Schaik et al. 1983b).

Finally, leopard presence itself was not a good estimate of predatory threat because leopards (1) often did not attempt to hunt primates that were nearby, especially vervets; (2) avoided hunting baboons during the day even when the baboons were close by, but instead hunted them at night; and (3) hunted vervets during the day but not at night, even though they were around vervets more often at night. Predator sympatry with primates, the number of predator species, and predator density have long been used as indirect measures of predation pressure supporting the claim that primates in more open habitats are at greater

Figure 6.5 Typical tall tree, riverine habitat of vervets at Mpala, in Laikipia, Kenya. Vervets may have been able to detect leopards more reliably than baboons, despite their smaller group sizes, because they had a better angle of view from the trees. Note the vervet at the top of the tree.

risk of predation than primates in forested habitats (DeVore 1963; Crook and Gartlan 1966; Clutton-Brock and Harvey 1977a; Terborgh and Janson 1986; Dunbar 1988). Such indirect measures assume that predators will always attempt to hunt primates if they are nearby, but the synchronous study of leopards, olive baboons, and vervets reveals the one-sided nature of that assumption. The motivation of the predator and its willingness to hunt at any moment in time is a better measure of, and is key to, assessing the likelihood of attack. If the predator is not motivated to hunt because it has just fed, it is searching for different prey, or it is traveling, for instance, there may be no actual immediate risk even when the predator is nearby, and prey often know this. For example, ungulates do not invariably flee when they see predators because they can assess risk by cueing in on the predator's location and behavior (Walther 1969). Only more such simultaneous observations of predators and their prey will give us a better sense of predation risk.

To summarize, many of the expectations of the predation hypothesis for group living were not upheld when primates and leopards, a major predator of catarrhine primates, were studied together. Whether within or between primate species, (1) larger groups were not better able than smaller groups to detect this important ambush predator; (2) larger groups were not better protected than smaller groups from predation; (3) predation rates were not lower for the species with larger body mass; and (4) predator presence was an unreliable measure of predation risk. Predator presence, predator density, or numbers of predator species are not useful metrics for estimating predation pressure, and relying on them for so long may have hindered our understanding of primate socioecology.

Chimpanzees and leopards are but two predators of primates, and they hunt in different ways. Consider, in addition, the diverse hunting styles of various raptors, snakes, and other carnivorans. As Zuberbühler and Jenny (2002:876) pointed out, "predation is often treated as a homogeneous evolutionary force, even though predators differ considerably in their hunting behaviour and the consequential selective pressures they impose on a primate community" (see also Treves 1999). Even the predatory behavior within a given predator species can be highly variable. Primates can adjust their behavior accordingly to some degree, but broad statements about the evolutionary influence of predation on primate behavior and social organization are unwarranted given the evidence we now have.

Primates Are Not Ungulates

In the absence of direct data, the predation hypothesis has been supported by parallels drawn between primates and other mammals, especially ungulates (Alexander 1974; Clutton-Brock and Harvey 1977a; Terborgh 1983; van Schaik and van Hooff 1983). Like primates, small ungulates tend to be nocturnal and live alone or in small groups in densely vegetated habitats, and large ungulates tend to be diurnal and live in large groups in more open habitats (Jarman 1974). But let us look more closely.

Ungulates are quintessential terrestrial prey animals, with non-behavioral adaptations that are clearly responses to predation pressure, such as elongated foot bones for running, laterally placed eyes, precocial neonates, and fast life histories. Shorter lifespans, earlier maturation, and faster reproductive output are hallmarks of fast life histories and are thought to be driven by high extrinsic mortality, e.g., mortality caused by predation and disease (Young 1981, 1992; Promislow and Harvey 1990; Stearns 1992; Charnov 1993; Charnov and Berrigan 1993; Reznick et al. 2001; Reznick and Bryant 2007; Isbell et al. 2009).

In comparison, primates are largely arboreal and, generally, cannot outrun their predators. They also have forward-facing eyes (which may or may not be related to predation pressure; Cartmill 1992; Sussman 1992; Isbell 2006, 2009), altricial neonates, and slower life histories. Longer lifespans in animals have been linked to arboreality, as birds, bats, and primates have longer lifespans relative to body mass than other animals (Austad and Fischer 1991; Kappeler et al. 2003; Shattuck and Williams 2010). More recent work, however, suggests that volancy (the ability to fly) favors long lifespans more so than arboreality. Among non-volant species, arboreality and fossoriality (burrowing), are also associated with long lifespans (Healy et al. 2014). Consistent with their faster life histories, few ungulates (tree hyraxes, *Dendrohyrax* spp. are exceptions) are arboreal, and none fly or burrow.

In keeping with primates' slow life-history patterns, they have lower mortality rates than most other mammals of similar body mass (Charnov 1993), including ungulates. The mean annual mortality rate was 0.09 for adult female chacma baboons in the Moremi Game Reserve, Botswana (Cheney et al. 2004), and 0.15 for adult female vervets at Segera in Laikipia, Kenya (Isbell et al. 2009). To put vervet mortality in the context of another important selective pressure (food), with a mean annual reproductive rate of 0.62 and a mean annual survival rate of 0.85 (Isbell et al. 2009), it was easier for adult females in this population to stay alive each year than to reproduce. In contrast, the mean annual mortality rate of adult warthogs, *Phacochoerus aethiopicus*, in Kruger National Park, South Africa, was 0.34 (Owen-Smith and Mason 2005). Most of the extrinsic mortality in ungulates arises from predation (Sinclair et al. 2003; Grange et al. 2004; cf. Hart and Sussman 2005). Even carnivores can have higher mortality rates than primates. For example, the mean annual mortality rate of adult wild dogs, *Lycaon pictus*, in Laikipia, was 0.29 (Woodroffe 2011). Predation from other carnivores was the most common natural cause of death (Woodroffe et al. 2007). These differential mortality rates are consistent with life-history theory. It appears that ungulates, with their faster life histories driven by extrinsic mortality largely caused by predation, are not good stand-ins for primates, with their slower life histories.

If that is not enough, here is another argument against ungulates as analogs for primates. While it may be the case that most individuals eventually die of predation in many primate species, if predation tends to occur relatively late in life, its strength as a selective pressure will be relatively weak. Although chacma baboons in Moremi expressed the typical mammalian pattern of higher infant than adult female mortality, and most adult females eventually did die of predation, most infants that died did so from infanticide, not predation (Cheney et al. 2004). Infanticide is a more intense selective pressure than predation because victims of infanticide clearly have no chance of reproducing, whereas adult females, who have obviously survived that vulnerable age, may have numerous opportunities to reproduce before they die of predation. Infanticide is far less common in ungulates than in primates (Lukas and Huchard 2014).

We have already seen that the process of predation is highly variable even for just one predator–prey relationship. By using ungulates as proxies for primates, we are continuing to treat predation as a simple unitary phenomenon, ignoring the great diversity of predators and their hunting styles, preferred habitats, motivations, and so on. What is the likelihood that similar patterns of body mass, diel activity, and social organizations are closely tied in both ungulates and primates to something as variable as the behavior of multiple kinds of predators with different predatory styles?

When we expand to other mammals, our treatment of predation pressure as a unitary phenomenon responsible for the pattern of small body mass = nocturnality = solitary foraging versus large body mass = diurnality = group living becomes even more problematic. For instance, treeshrews, *Tupaia* spp., which can be as small as 50 g (Emmons 2000), rock and tree hyraxes, *Procavia johnstoni* and *Dendrohyrax brucei*, respectively, dwarf and banded mongooses, *Helogale parvula* and *Mungos mungo*, respectively, and African striped and South African ground squirrels, *Xerus erythropus* and *Xerus inauris,* respectively, are all diurnal (Turner and Watson 1965; Rasa 1987; Linn and Key 1996; Emmons 2000; Gilchrist 2006; Barocas et al. 2011; Thorington et al. 2012) despite experiencing heavy predation (Rasa 1987; Barry and Barry 1996; McPherson et al. 2016; Naude et al. 2019). I have often seen small (~80 g; Packer 1983) Nile grass rats, *Arvicanthis niloticus*, in East African savannahs scurrying about during the day (see also Blanchong and Smale 2000). Many of these species also live in extended families or groups (Turner and Watson 1965; Rasa 1987; Linn and Key 1996; Gilchrist 2006; Barocas et al. 2011). If predation favors nocturnal solitary foraging, why are they diurnal, and why do many of them live in groups?

Main Points

When predation is approached from several angles, it does not hold up well as the main selective pressure favoring nocturnal solitary foraging and diurnal group living. Among the criticisms are that (1) predation's execution is too variable to be a unitary pressure, (2) nocturnal primates are not very cryptic, (3) empirical evidence from primate interactions with leopards does not support long-standing assumptions of the predation hypothesis, and (4) life history theory does not support that predation pressure is responsible for convergences in diel activities and social organizations in primates and other mammals. We need to look for alternatives.

CHAPTER 7

Resolving the Nocturnal/Diurnal and Solitary/Group Forager Divides

Our findings demonstrate that biogeographical patterns
of global mammalian diversity are structured in part by
the availability of the temporal niche, which is itself
constrained by natural cycles of both light and
temperature.

— J. J. Bennie et al. (2014:13729–30)

There must be a selective pressure more universal and consistent than the highly variable selective pressure of "predation" that can explain better the convergences in body mass, diel activity, and social organization that exist among primates, ungulates, and other mammals. In this chapter, I will first discuss the link between body mass and the temporal niche and then segue into the link between the temporal niche and solitary foraging versus group living in primates. As it is well known that small body mass is associated with nocturnality, and large body mass, with diurnality in mammals, there must be something universal and consistent about the diel period that favors this dichotomy regardless of where in the world mammals live. Of course, what immediately comes to mind is a difference in sunlight, which is what the predation hypothesis emphasizes, but sunlight affects air temperature, too.

Thermal Constraints: An Alternative Hypothesis
for the Nocturnal/Diurnal Dichotomy

The thermal environment, which combines air temperature, solar radiation, wind speed, and relative humidity to affect heat load in animals (Tracy and Christian 1986), is in several ways a very consistent selective pressure. It has always existed, is warmer with sunlight than without, and is less variable daily and seasonally in the tropics than in temperate zones. Different animals have adapted to this evolutionary reality in different ways. For instance, artiodactyls (a subset of ungulates)—but not primates—selectively cool their brains anatomically and physiologically via the internal carotid rete, a network of arterioles stemming from the carotid artery, thereby reducing the risk of hyperthermia and dehydration under warm, dry conditions (Maloney et al. 2007; Strauss et al. 2017). (Lorisoids, e.g., galagos and lorises, actually have a simple external carotid rete; Kanagasuntheram and Krishnamurti 1965; Butler 1980.) The difference between artiodactyls and primates likely reflects both phylogeny and their different ecological niches. Diversification of artiodactyls increased as their habitats became warmer and drier (Strauss et al. 2017), whereas primates largely diversified in wetter tropical forests, where the daytime thermal environment is modulated by canopy shade and water is more readily available.

Across mammals, air temperature is a strong predictor of diel activity at both ends of the temperature spectrum (Bennie et al. 2014). In cold habitats, small mammals such as squirrels, chipmunks, and other rodents are more likely to be diurnal than nocturnal as they take advantage of the warmer temperatures during the day (Roll et al. 2006; Guiden and Orrock 2020). In warmer habitats, small mammals are more likely to be nocturnal—and large mammals, diurnal—because small mammals are at greater risk of hyperthermia and death under high temperatures (Fuller et al. 2021). Hot thermal environments are more problematic for small mammals partly because they have higher surface area to volume ratios than large mammals, making internal and body surface temperatures more difficult to regulate (Phillips and Heath 1995; Fuller et al. 2021). Small mammals thus rely less on maintaining consistent body temperature than on adjusting body temperature and metabolic rate in response to cold and heat, e.g., through torpor (Lovegrove et al. 2014; Blanco et al. 2018).

In general, basal metabolic rate decreases with decreasing body mass and increasing ambient temperature (Careau et al. 2007; Clarke et al. 2010), but ambient temperature more strongly affects basal metabolic rate in smaller mammals than in larger mammals. This is possibly because, besides having lower surface area to volume ratios, larger mammals have fewer other constraints that affect thermal balance (Naya et al. 2018). If one wants to minimize the risk of hyperthermia in a warm environment, being small and nocturnal with a low basal metabolic rate would be one way to do it. Given that primates evolved not in the temperate zones but in the warmer tropics, nocturnality in smaller primates and diurnality in larger primates might be better explained as an evolutionary response to thermal constraints than as an anti-predator strategy.

The smallest primates, some weighing as little as 30 g, can be found among the nocturnal cheirogaleids (mouse and dwarf lemurs) of Madagascar (Masters et al. 2014). They have undergone phyletic dwarfism, an evolutionary process whereby body mass is reduced over time (Ford 1980; Martin 1992; Montgomery and Mundy 2013; Masters et al. 2014). It has been suggested that phyletic dwarfism is an evolutionary response to the unpredictable but recurrent harsh climatic conditions, especially severe droughts, brought on by El Niño–Southern Oscillation (ENSO) events (Génin et al. 2010; Masters et al. 2014). Droughts in warmer habitats decrease the availability of water and increase the likelihood of dehydration and heat stress (Fuller et al. 2021). Mouse and dwarf lemurs also exhibit heterothermy, in which body temperatures temporarily match ambient temperatures and basal metabolic rates decrease during resting periods (Blanco et al. 2018). Golden brown mouse lemurs and gray-brown mouse lemurs, *Microcebus griseorufus*, had higher skin or body temperatures during the day when they rested than during the night when they were active (Lovegrove et al. 2014). Although heterothermy is considered an energy-saving adaptation (Masters et al. 2014; Blanco et al. 2018), heterothermic mammals might also be reducing the risk of hyperthermia in warm and drought-prone environments.

Nocturnality and heterothermy may be valuable ways for small mammals in the tropics to avoid heat stress (Levy et al. 2019; Bonebrake et al. 2020), but there are, nonetheless, small diurnal mammals in warm, tropical habitats, including hyraxes, mongooses, squirrels, and callitrichine primates. For example, the pygmy marmoset, *Cebuella pygmaea*, the smallest monkey, weighs roughly only 150 g (Genoud et al. 1997). The predation hypothesis does not attempt to explain their diurnality, and they also seem anomalous from the perspective of thermal constraints. If thermal constraints determine active diel periods in primates, as I am proposing, why did pygmy marmosets not evolve nocturnality to avoid hyperthermia, and how do they avoid overheating during the day? I suggest that the answer to both questions lies in the difficulty of switching from diurnality to nocturnality. For primates, especially, switching would require major changes to the visual system. The difficulty of becoming secondarily nocturnal is suggested by the rarity of switching: only tarsiers and owl monkeys and perhaps woolly lemurs, *Avahi* spp., appear to have made the switch from diurnality to nocturnality (Müller and Thalmann 2000; Williams et al. 2010; Koga et al. 2017). That change included evolving very large eyes and, in the case of owl monkeys, losing color vision (Jacobs et al. 1993; Williams et al. 2010). I am not saying that tarsiers and owl monkeys became nocturnal to avoid heat stress, although, why not? After all, they are small primates with basal metabolic rates lower than expected for their body mass (Le Malo et al. 1981; McNab and Wright 1987; Genoud et al. 1997; Genoud 2002). In fact, one study showed that tarsiers had maximum skin temperatures 2°–3° C *higher* during the day when they rested than during the night when they were active, and when ambient daytime temperatures were higher, skin temperatures during the day were also higher (Lovegrove et al. 2014), even though tarsiers are not heterothermic (Welman et al. 2017). Just imagine how high their skin temperatures might go if these very small primates (females weigh 52–151 g; references in Rowe and Myers 2016) were to become active during the day. For small primates that could not cross the Rubicon from diurnality to nocturnality (cathemeral primates—those active at irregular intervals during the day or night—might be viewed as living on a small island in the Rubicon waters), an alternative evolutionary path in warm climates might have been phyletic dwarfism. Like cheirogaleids, pygmy marmosets (and other callitrichines) have undergone dwarfism, and they have among the lowest basal metabolic rates relative to body mass of any anthropoid primate yet measured (Genoud et al. 1997; Genoud 2002), which might help to keep their body temperatures from rising too high during the day.

Diurnal mammals in warm climates can also protect themselves from hyperthermia by making behavioral adjustments. For example, rock hyraxes move to the undergrowth near kopjes (Turner and Watson 1965); meerkats, *Suricata suricatta*, retreat to burrows (Doolan and Macdonald 1996); and African striped ground squirrels seek shade, use their tails as umbrellas, or rest splayed out flat ("splooting" or "hearth-rugging") on the ground (Linn and Key 1996; Thorington et al. 2012). I have recently even seen invasive fox squirrels, *Sciurus niger*, splooting in my so-called "temperate zone" California backyard during record-breaking summer heat waves.

Tropical forests are cooler at lower levels (Allee 1926; Thompson et al. 2016), and we would expect smaller primates to spend their time at the lower levels if they are at greater risk of hyperthermia. First sightings of seven primate species in Surinam showed that the three smallest species (492–1871 g) were indeed found at lower heights in the forest, and the two largest species (7275–7775 g), at greater heights (Fleagle and Mittermeier 1980). Similarly, a survey of 12 primate species in Bolivia showed that all but one, a callitrichine, were first sighted at lower heights, on average, than the three largest species (Buchanan-Smith et al. 2000). In another study, pygmy marmosets spent their time at the lowest heights when they traveled (mean: 3.2 m) and at the highest when they rested (mean: 7.3 m) (Youlatos 2009), differences that also make sense from a thermoregulatory perspective.

Ring-tailed lemurs shift to cathemerality as ambient temperatures increase (Donati et al. 2011). When ambient temperatures are high, Verreaux's sifakas sit on the ground, wrapping their arms around the trunks of trees and pressing their bodies against the trunks, which are cooler than both the air temperature and locations higher up on the trunks

(Chen-Kraus et al. 2023). Red-fronted lemurs and white-faced capuchins rest more during the hottest times of the day during the hot dry seasons (Campos and Fedigan 2009; Sato 2012). Prince Bernard's titi monkeys, *Plecturocebus* (formerly *Callicebus*) *bernhardi*, avoid sunny locations in forests and rest in body positions that dissipate heat (Lopes and Bicca-Marques 2017).

In more open habitats, chacma baboons seek shade and become more sedentary during the heat of the day (Hill 2006). Drinking water lowers their body temperature fairly quickly (Brain and Mitchell 1999). It has also been suggested that their "sandbathing," which involves tossing subsurface sand at the ventrum, is a thermoregulatory behavior (Brain and Mitchell 1999). Vervets also seek shade, and their maximum body temperatures decrease when they spend more time in the shade (McFarland et al. 2020). However, primates are not always successful at avoiding heat stress. Although chimpanzees living in Fongoli, a savanna-woodland environment, go into caves and pools of water during hot periods (Pruetz 2007; Wessling et al. 2018), cortisol levels, as a measure of stress, and creatinine levels, as a measure of dehydration, indicate that they still experience heat stress and dehydration during these hot times (Wessling et al. 2018).

Recall that fossorial mammals, flying birds and bats, and primates all have longer lifespans than expected for their body size (Healy et al. 2014). This convergence of such disparate groups was attributed to low extrinsic mortality, due particularly to low predation (Healy et al. 2014). I question low predation as a driver of their extraordinary longevity given that mortality from predation is highly variable among primates alone and that the variation would almost certainly increase when the non-primate taxa are added. What might take its place? I suggest that heat stress as a major source of extrinsic mortality, especially in warmer climates, might explain the convergence better than predation. What all these animals have in common besides their longevity is their ability to live in or retreat quickly to cooler microclimates, such as caves, subterranean burrows, bodies of water, tree holes, and trees (see also Mc-Cain and King 2014). The thermal environment thus may have been a selective pressure strong enough to have molded life histories, not just the nocturnal/diurnal divide, and I suspect its importance will become more obvious as our planet continues to heat up.

Foraging Efficiency Mediated by Sunlight May Explain the Dichotomy between Diurnal Group Living and Nocturnal Solitary Foraging

Consistent with the hypothesis presented in this book that foraging efficiency is the ultimate driving force behind (1) the extent to which females can expand their home ranges to include additional reproductive females, (2) female movement strategies, and (3) female social organizations, I maintain that foraging efficiency was also the impetus for the evolution of group living in primates. A female who shares her home range by expanding it enough to accommodate other reproductive females likely has enough food for her own reproduction under normal conditions. Otherwise, the practice of sharing the home range would be too costly and would not have been favored evolutionarily. Nonetheless, even with all the land in the world, a female's foraging efficiency would suffer if another individual were to arrive at a food site first and eat all the food there. That female will have wasted the energy required to go to that food site and now must expend more energy to go elsewhere, with no guarantee that the loss will not happen again. She can minimize that inefficiency, however, by making sure she knows where the others in her home range have eaten recently and avoiding following in their footsteps as she feeds. This applies equally to diurnal and nocturnal primates. How might females monitor the movements of others in a shared or partially shared home range? The answer to that question is different for diurnal and nocturnal females.

As vision is the main sensory system in diurnal primates, it is reasonable to expect that they would use vision to monitor the locations of others, but there are limitations—visual monitoring is more difficult with greater distance and denser vegetation. Spatial coalescing makes visual monitoring easier (Suzuki and Sugiura 2011; Sugiura et al. 2013). Thus grouping, i.e., females drawing together and consistently traveling near each other, may be seen as a consequence of the ability to visually monitor others in a shared home range in order to maintain individual foraging efficiency. Vocalizations can also be used for locational monitoring (e.g., Lima et al. 2019), but they appear to be secondary to vision among diurnal primates, as vocalization rates within groups have been reported to increase when visibility is reduced (Caine and Stevens 1990; Koda et al. 2008). Other complexities and benefits of group living, such as nuanced social relationships among females (Silk 2012), would have evolved after this incipient stage of group living. These more complex aspects of group living would be a step away from ecology and determined more by social factors. Viewing group living from that perspective might help us understand the success of socioecology in explaining elemental issues such as those presented in this book and its limited success with more fine-grained variation in social behavior and relationships (Ménard 2004; Balasubramaniam et al. 2012a, b; Thierry 2008, 2013). Indeed, this incipient stage of group living might be better called "group foraging" rather than "group living" to distinguish its functions and benefits from the social functions and benefits that would have developed later.

Visual monitoring as a way to maintain individual foraging efficiency would not reliably apply to nocturnal primates. Whereas sunlight is rarely filtered enough to create darkness during the day, moonlight goes through different phases of illumination every month such that some nights are very dark and others are very light. Many nocturnal primates are sensitive to this variation and become more active when the moon's illumination is strong, suggesting that they do make use of their vision to navigate when possible (Gursky 2003; Bearder et al. 2006; Fernandez-Duque and Erkert 2006). However, since nocturnal primates cannot consistently use sunlight or moonlight to monitor others visually, there is no advantage to coalescing into groups to maintain foraging efficiency in shared home ranges. Against the background of variable illumination, nocturnal primates use their auditory and olfactory senses more than do diurnal primates (Eisenberg et al. 1972; Ramanankirahina et al. 2016; Drea et al. 2019). Vocal signals have the farthest spatial reach, while olfactory signals have the longest temporal reach (Ramsier and Rauschecker 2017). In other words, vocal communication can assist individuals in determining where others are at a moment in time, and olfactory communication can assist them in determining when others were present in a particular place. Importantly, unlike visual signals, vocal and olfactory signals do not require individuals to remain in close spatial proximity.

Vocal and olfactory signals in nocturnal primates have been interpreted in general as a spacing mechanism to avoid agonistic interactions (Braune et al. 2005; Bearder et al. 2006). In several species of galagos, individuals also vocalize back and forth with each other throughout the active period. These have been interpreted as contact or cohesion calls (Bearder et al. 2003). Spacing and cohesion call functions are not only compatible with the function of avoiding the feeding locations of others, but they also reinforce each other. Red-tailed sportive lemurs, *Lepilemur ruficaudatus*, vocalize regularly while active, even outside of social interactions, leading Fichtel and Hilgartner (2013) to suggest that the vocalizations indeed function to identify locations to partners as well as neighbors in different home ranges. Olfactory communication is still poorly studied (Colquhoun 2011). However, "urine washing", defined as the act of rubbing urine onto the feet by the hands and then deposited where individuals walk (Charles-Dominique 1977), has been studied and is a good candidate example of how some nocturnal primates use olfaction to identify where and how long ago others have been. For example, female Allen's galagos sometimes share their home ranges extensively with their adult daughters, and they mark their paths by urine washing throughout their home ranges, which could be done to identify and avoid locations where their relatives have fed recently. They mark most often in areas of overlap with other apparently unrelated females, suggesting that the behavior also has a territorial function (Charles-Dominique 1977). As with spacing and cohesion vocalizations, urine washing is fully compatible with both territoriality and the function I am proposing, that of avoiding locations where others have recently fed to maintain foraging efficiency.

How to Explain the Diurnal, Solitary-Foraging, Female Orangutan Social Organization?

While there are no nocturnal, group-foraging primates, there is one diurnal, solitarily foraging primate. The exceptional nature of the orangutan's social organization has prompted several attempts at explanation. One is that orangutans were more social in the past but that *Homo erectus*, and later, *Homo sapiens*, hunted them so relentlessly that they shifted to greater arboreality, solitary living, and crypsis to avoid hominins (MacKinnon 1974; Jolly 1999). Another is that there is no need to group because they are too big or too arboreal to be vulnerable to heavy predation (van Schaik and van Hooff 1983; Mitani 1989; Kunz et al. 2021). A third is that their foods keep them from forming groups because individuals rapidly deplete the foods found in trees (Galdikas 1988). These latter two appear to pit predation against food, but both are actually consistent with the most accepted socioecological model that predation forces diurnal primates to live in groups (which I disputed in chapter 6), while food competition sets an upper limit to the size of groups (van Schaik and van Hooff 1983; Janson and Goldsmith 1995; Sterck et al. 1997). Let me propose instead an explanation that is consistent with the Variable Home Range Sharing model's emphasis on foraging efficiency.

The fact that female orangutans share their home range to some extent with other females indicates that expanding their home range to accommodate other females is not prohibitive. However, the risk of reduced foraging efficiency is increased in shared areas when others feed there first, so they need to manage this somehow. It does not appear that orangutans are indifferent to where others feed; in a captive experimental study, orangutans avoided going to food sites that they saw another orangutan deplete (MacDonald and Agnes 1999). The question from the perspective of the Variable Home Range Sharing model's focus on foraging efficiency is, why have they not taken advantage of sunlight to visually monitor the movements of others in shared parts of their home range? In fact, I believe they have.

As the largest habitually arboreal mammal (Thorpe et al. 2009), orangutans have unique characteristics that should make it possible for them to visually monitor the locations of others from greater distances than other diurnal primates can do. Any fieldworker knows that larger-bodied individuals are easier to see than smaller-bodied individuals at greater distances, all else being equal. Since orangutans are very large-bodied, they should be able to see each other from farther distances than can conspecifics in other species. How far can they see in the forest? Cheryl Knott and colleagues (2008) used a distance of 50 m to operationally define social encounters between female orangutans, suggesting that the orangutans were aware of each other at that distance, but this measure was based on how far humans on the ground can see. When they climbed up into the canopy, they confirmed that up there humans can see even farther than 50 m, and they suggested that orangutans in the canopy can see farther as well.

Another characteristic that may make orangutans visible from farther away in dense vegetation is that they sometimes use their large bodies to sway branches and trunks as they move through the canopy (MacKinnon 1974; Cant 1987; Thorpe et al. 2007, 2009). Thus, even if they cannot actually see each other, their locations can sometimes be identified from the swaying vegetation.

While both diurnal and nocturnal primates can maintain their foraging efficiency by avoiding where others have already fed, their processes for doing so are different and have different consequences. Diurnal primates use visual monitoring to avoid where others have already fed, resulting in the formation of groups as individuals draw closer together to see where others are going and feeding. Since vision at night is less useful than other senses for conspecific monitoring, nocturnal primates instead avoid where others have already fed by spacing themselves out. Orangutans are a cross between diurnal and nocturnal primates in how their monitoring works. Like other diurnal primates, they monitor others visually, but over longer distances. This allows them to maintain foraging efficiency by avoiding where others have been, but it also helps give us the impression that they do not live in groups because they do not coalesce as compactly as other diurnal primates.

On occasion, they do move closer together when proximity does not reduce foraging efficiency, such as when foods are locally highly abundant (Sugardjito et al. 1987; see also Mitani et al. 1991). After feeding together, they often then leave and travel together for some time within 30 m of each other (Sugardjito et al. 1987), a distance which, again, should be in their visual range (Mitani et al. 1991). This (and other evidence) has even led some researchers to argue that some populations of orangutans are better described as having a fission-fusion society (van Schaik 1999; Singleton and van Schaik 2002). Recognition of their unique ability to monitor each other visually at far distances may help us better understand their unique combination of diurnality, solitary foraging, and fission-fusion ranging.

★ ★ ★

Throughout this book, I have purposely limited discussion of social organizations to females because they are the "ecological sex," but primate social organizations are not complete until we add in the males. In the next chapter, we will focus on the selective pressures operating on males and how they combine with those operating on females to create the full picture of primate social organizations.

Main Points

Similar social organizations in primates and other mammals that have been attributed to predation could have been driven by other selective pressures shared in common. I propose that those selective pressures are thermal constraints for the nocturnal/diurnal divide, and foraging efficiency mediated by the absence or presence of sunlight for the solitary/group forager dichotomy. Admittedly, the thermal environment and sunlight are not as "flashy" as predation, but they have been far more consistent and reliable, providing the opportunity for directional selection to influence female primate social organizations. Thermal constraints on body mass determine diel activity, and in warmer tropical environments where primates evolved, small-bodied mammals tend to be nocturnal, and large-bodied mammals, diurnal—including primates. Diurnality makes it possible for females to monitor others visually as a way to avoid where those others have fed, thereby maintaining their foraging efficiency in shared parts of the home range. Incipient groups would have formed when some primates became diurnal because visual monitoring requires fairly close spatial proximity (except for orangutans). After that step, other benefits of grouping, including those that close social relationships manifest, would have formed largely independent of ecology. In contrast, because nocturnality precludes the use of sunlight to visually monitor others in shared parts of the home range, nocturnal primates monitor each other through vocal and olfactory cues, which do not require close spatial proximity. For them, foraging efficiency is generally best maintained in shared parts of the home range by staying away from each other. Finally, it is important to stress that, even if the Variable Home Range Sharing model does not ultimately hold up, disproving it would not affect the hypotheses that thermal constraints and sense-driven monitoring have affected diel activity cycles, solitary foraging, and group foraging.

CHAPTER 8

Male Contributions to Female Social Organizations

In brief, the adaptive bases for pair-living are far from
being completely understood.
—P. M. Kappeler and C. P. van Schaik (2002:715)

Social organizations are at their heart descriptions of variation in space use because the ways in which animals use their space determine with whom they come into contact. Now that we have covered female social organizations as the core basis for primate social organizations, it is time to bring males into the discussion. There is little dispute that males have been under strong selection to seek out females as mating partners because females limit male reproductive success (Darwin 1871; Goss-Custard et al. 1972; Trivers 1972; Altmann 1990; Mitani et al. 1996); however, as the epigraph for this chapter tells us, how that behavior ends up affecting female social organizations to create primate social organizations is not completely resolved. In this chapter, I will summarize current views first. I will then examine male space use from the perspective of movement strategies to see if this approach, when combined with selection to seek out females, can explain unusual male contributions to female social organizations that are still unresolved, i.e., pair living and supernumerary males in small groups of females. Finally, I will discuss the two female social organizations for which predictions about female movement strategies in relation to home range sharing cannot be made either because males control female home range boundaries or because home ranges are chosen by females more for non-food factors—such as reducing predation, infanticide, or the risk of inbreeding—than for food. In all these situations, males play a prominent role. Because spacing systems ultimately determine social organizations, please note that I am only concerned with spacing systems here, not mating systems.

Current Views on Male Space Use

Early on, Crook and Gartlan (1966) recognized that adult male membership in female groups tends to increase as the numbers of adult females in those groups increase. This observation has been reinforced over the years by many studies, and the pattern is easily explained as a consequence of sexual selection: males attempt to gain proximity to as many potential mating partners as possible, with poorer success in monopolizing access as the number of females in proximity increases (e.g., Kummer 1968; Eisenberg et al. 1972; Goss-Custard et al. 1972; Emlen and Oring 1977; Harcourt 1978; van Schaik and van Hooff 1983; Terborgh and Janson 1986; Altmann 1990; Mitani et al. 1996). For example, when adult female cercopithecine primates travel in groups of up to five, they are typically accompanied by one adult male who travels and feeds with them in the group's home range, whereas groups of more than 10 adult females are accompanied by multiple males. Groups of between 5 and 10 females may be accompanied by either one or more than one male (Andelman 1986). Maximizing access to potential mating partners can also explain male home ranges that overlap those of multiple adult females when females have their own home ranges (Clutton-Brock 1989; Singleton and van Schaik 2002; Schmelting et al. 2007).

There are two apparent exceptions to this "rule" of maximizing access to potential mating partners. One is when females have their own home ranges and male home ranges overlap only with that of only one female, i.e., pair living in the traditional sense. Why would a male configure his space use to gain access to only one female? The other is when there are "extra" males in small groups of females. Why does a single male not exclude all other males from such small groups?

Attempts to answer these questions are usually couched in terms of an inability of males to defend larger home ranges (the resource defense hypothesis; Ellefson 1968; Emlen and Oring 1977) or multiple females (the female defense hypothesis), especially at low female densities (Emlen and Oring 1977; Lukas and Clutton-Brock 2013). There may also be benefits, however. Suggested benefits when a male shares the same home range with only one female include increased paternity certainty, which sometimes leads to paternal care and greater offspring survival (Clutton-Brock and Harvey 1977a; Clutton-Brock 1989); risk minimization, e.g., from predation, intersexual aggression, or energy costs from travel (Hilgartner et al. 2012); and protection of offspring from infanticidal males (van Schaik and Dunbar 1990; Borries et al. 2011; Opie et al. 2013). Protection against infanticide has also been proposed for "extra" males in small female groups (van Schaik 1996; Crockett and Janson 2000; Treves 2001; Port et al. 2010). Stronger home range or group defense is another suggested benefit of the presence of supernumerary males in small female groups (Port et al. 2010).

Yet Another Potential Explanation for Pair Living

Perhaps we are so focused on male access to females that we forget the fact that males also need to eat well if they are to be effective in competing against other males for access to females. Unlike females, who have been selected to maximize their efforts to obtain food for successful reproduction, males have had to respond to two very different but related selective pressures—they need to maximize their efforts to obtain both food and females in order to reproduce successfully. Males should thus benefit from moving in ways that improve or at least maintain their foraging efficiency because they can then devote more energy toward developing competitive body condition and engaging in behaviors more directly involved in mate acquisition. Since male and female conspecifics living in the same area generally eat the same foods if not always in the same proportions (e.g., Chivers 1977; Rose 1994; Isbell 1998; Schülke 2003), it is probably safe to assume that males employ the same movement strategies that female conspecifics employ with respect to food; those that do not probably would not be able to travel consistently with females (e.g., gray-cheeked mangabeys).

One result of selection on males to seek out mating partners is that, with few exceptions, males disperse from the natal group or home range. Recall that the directed travel movement strategy requires knowledge of the temporal and spatial locations of foods. Like their female counterparts, dispersing males who mainly employ the directed travel movement strategy are expected to suffer a reduction in foraging efficiency from a lack of knowledge of the locations of foods when they move into new areas, all else being equal. Foraging inefficiency could be avoided, however, if they find and travel with someone who already knows the area. The strategy of using others to identify suitable habitat has been observed in a variety of taxa. Duck hunters have long known that ducks are drawn in by conspecifics, and they exploit that fact by using decoys (Reed and Dobson 1993). Scientists have also begun to exploit conspecific cueing of good habitats to conserve populations. One recent study used visual and acoustic "decoys" (feces and vocalizations) of burrowing owls, *Athene cunicularia*, to convince translocated owls to stay in a new habitat (Hennessy et al. 2022). Male orangutans dispersing to unfamiliar areas watch locals intensely (called "peering") most often when the locals are eating foods, and peering declines over time as the dispersers become more familiar with the area (Mörchen et al. 2023). Similarly, a dispersing male primate who employs mainly directed travel might use conspecific cueing to identify a good habitat and to learn where and even what foods are from a knowledgeable female in her home range. He might initially tail the experienced female as she goes from food site to food site. Once the male becomes familiar with the home range, he might take the lead more often to food sites or travel less often with the female. Females that employ mainly directed travel are expected to benefit in the same way when they disperse by finding and moving with a knowledgeable male in his home range, at least until they learn the new home range. Established pale fork-marked lemur pair-partners stayed about 100 m from each other on average (Schülke and Kappeler 2003), which is farther apart than observed in some other pair-living species (e.g., Palombit 1994;

Dolotovskaya et al. 2020). It would be interesting to know more about the movements and foraging efficiency of immigrant fork-marked lemurs. Do new immigrants initially stay closer to the more knowledgeable individual, or do they remain farther away and suffer reduced foraging efficiency compared to the knowledgeable individual until they have spent more time in that home range?

The female counterparts of males who are predicted to use the directed travel strategy most are Isolationists. Focusing on movement strategies as the mechanism and foraging efficiency as a selective pressure, Isolationists are expected to form pairs since male directed travelers who disperse maintain their foraging efficiency best by finding and traveling with females, and Isolationist females do not share their home range with other adult females. This hypothesis for pair living is consistent with the finding that, across mammals, pair living is associated with unshared female home ranges (Komers and Brotherton 1997).

In contrast, when dispersing males mainly employ the opportunistic searching movement strategy, their foraging efficiency is not expected to decline as much when they move into new areas because searching for and often failing to find food are defining characteristics of that movement strategy. Such males may not need to travel with females who know the area in order to maintain their foraging efficiency, and that freedom should allow them to increase their access to more females by not restricting their home range to just one female. Their female counterparts include Generous females, who share home ranges to some extent with other females. I suggest that, ultimately, differences in male movement strategies may explain why in some species in which females have their own home ranges, e.g., titi monkeys and siamangs, one male and one female typically share the same home range, while in other species, e.g., mouse lemurs and orangutans, male home ranges are larger than that of any one female (Figure 8.1).

Supernumerary Males in Constraining Female Social Organizations

In many primate species, groups tend to include two or more adult females per adult male (Clutton-Brock and Harvey 1977b; Mitani et al.

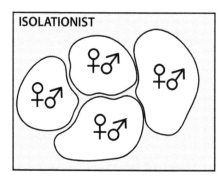
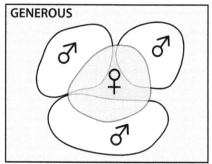

Figure 8.1 Male movement strategies typically mirror those of females. Extending the Variable Home Range Sharing model to male movements, pair living occurs because dispersing individuals employing mainly directed travel (Isolationists) benefit most by finding and traveling with someone of the opposite sex who is knowledgeable about the home range. In contrast, when dispersing males employ more opportunistic travel, they are not constrained to find and stay in the same home range with a knowledgeable female and are able to overlap their home ranges partially with those of multiple females, which maximizes their reproductive success.

1996; Altmann 2000). This is not often the case, however, for several species of lemurs (Kappeler 2000; Pochron and Wright 2003), howler monkeys (Crockett and Eisenberg 1987; Van Belle and Estrada 2008), and colobines (Sterck and van Hooff 2000) where adult males sometimes even outnumber adult females within groups. A bias toward adult males in groups is difficult to explain with sexual selection theory. In these species, the number of females per male tends to be smaller and more fluid, resulting in considerable variation in adult compositions among groups, even in the same population. At low densities, for instance, pairs may be common, but as populations grow, groups may develop into both single-male, multi-female groups and multi-male, multi-female groups, which could be age-graded (Eisenberg et al. 1972; Kappeler 2000; Pope 2000; Sterck and van Hooff 2000; Ostro et al. 2001; Teichroeb et al. 2009).

It may not be a coincidence that both the presence of "supernumerary" males and the extent of variation in the number of males in female groups are most often found in Constraining female social organizations. Constraining females are expected to have a somewhat limited ability to expand their home ranges, resulting in poorer reproduction, targeted aggression, and eviction of females, e.g., red-fronted lemurs (Kappeler and Fichtel 2012, 2016), when they reach the membership size limit that their home range can support. The Constraining female social organization often includes agonistic intergroup relationships along with male involvement in home range or group defense (e.g., Sekulic 1982b; van Schaik et al. 1992; Fashing 2001a; Sicotte and McIntosh 2004; Benadi et al. 2008; Gibson and Koenig 2012). Male involvement in group defense among lemurs, howlers, and colobines is usually couched in terms of defending groups against incursions by competitors or infanticidal males (van Schaik et al. 1992; van Schaik 1996; Crockett and Janson 2000; Pope 2000; Van Belle and Estrada 2008; Port et al. 2010; Kappeler and Fichtel 2012). In an investigation of 26 populations of five species of howler monkeys, Adrian Treves (2001) found that having more males in groups improved female reproductive success, which he suggested could be due to male defense of infants or food resources. Although supernumerary males might reduce food availability for females in groups, this cost may be outweighed by the benefits of male defensive behavior.

While this may explain the presence of supernumerary males in groups of Constraining females, the same benefits should apply to Promoting females because they also stand to gain from male defense of the group, food, or infants. Why do groups of Promoting females not then include more than the expected number of males? We need to examine other possibilities for the presence of supernumerary males in Constraining female groups that would not also apply to Promoting females.

One possibility is related to their different dispersal patterns. While Promoting females rarely disperse, Constraining females often do. It is not often easy for dispersing Constraining females to establish a new home range or join other groups, but male assistance appears to make it easier. For example, evicted female red-fronted lemurs were only able to establish a new home range if they attracted adult males (Kappeler and Fichtel 2012), and female red howlers were unable to begin reproducing until they found at least one male with whom to associate and establish a home range (Pope 2000). Dispersing female ursine colobus preferred to join groups whose males were successful in intergroup encounters against the female's former group (Teichroeb et al. 2009).

As groups grow, having "extra" males in the group also appears to help females hold onto their home range. In Kirindy Forest, the number of males in Verreaux's sifaka groups—either in total or in addition to the dominant male—did not affect infant survival or the probability of group takeovers, but groups with the larger number of subordinate males won more aggressive intergroup encounters (Kappeler et al. 2009). Females actively encouraged subordinate males to join their group by maintaining proximity to them and responding to their vocalizations more than they did for the dominant male (Lewis 2008).

In addition, while Promoting females (most of which are cercopithecines) are often actively involved in aggressive intergroup encounters (Wrangham 1980; Isbell 1991; Cheney 1992; Cords 2002), Constraining females only rarely take part (van Schaik et al. 1992; Fashing 2011; but see Crockett and Pope 1993). Little female involvement in aggressive intergroup encounters has been interpreted as evidence against female resource competition (van Schaik et al. 1992). However, if supernumerary males in female groups are effective in home range defense, females may not need to participate very much (Fashing 2001a; 2011). In red howlers, groups with more adult and subadult males experienced lower rates of incursions than groups with fewer males (Crockett and Janson 2000). One might argue that female participation would only strengthen the effectiveness of home range defense, but females may not be able to participate regularly without a cost. Low energy levels appear to be common among identified Constraining females (e.g., Sekulic 1982b; Dasilva 1992; Fashing 2011), and perhaps their energetic limitations contribute to their reliance on male assistance in acquiring and maintaining home ranges (Figure 8.2).

Since diurnal females sharing the same home range travel together to maintain foraging efficiency through visual monitoring, males should benefit by traveling with them for the same reason regardless of the number of males in female groups. Visual monitoring is also useful for males to maintain access to females and deny other males access to them.

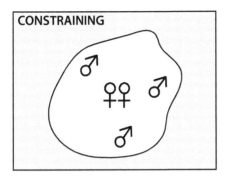
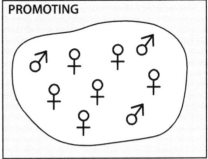

Figure 8.2 Male contributions to cohesive groups of females. In Constraining female groups, males may sometimes outnumber females. Females may encourage male membership, and males may help females gain and keep a home range. In Promoting female groups, since females seldom disperse and thus have little need for male help in acquiring a home range, male numbers are more aligned with expectations from sexual selection theory. In both types of females, males travel with the group because they also benefit from visual monitoring to maintain foraging efficiency.

Male Behavior and Bullied Females

While males help Constraining females establish and maintain their home ranges, there have been no reports of males actually controlling the size of Constraining female home ranges by aggressively patrolling home range boundaries and behaving aggressively toward females. Bullied females are those whose home range boundaries are determined by males engaging in such behaviors. Chimpanzees in East Africa are an excellent example of Bullied females, as articulated so well by Jennifer Williams and colleagues (2002:358): "In the absence of males, the best pattern for females in regard to feeding efficiency might be to space themselves evenly over the available habitat such that they can feed alone in areas where they have intimate knowledge of food distribution. However, the territorial behaviour of males renders boundary areas unsafe to most females, and appears to force them to cluster well within the defended range."

Female eastern chimpanzees are described as having core areas that overlap with those of other females (Wrangham 1979; Hasegawa 1990; Williams et al. 2002; Emery Thompson et al. 2007) (not unlike the home ranges of Generous females), and it is reasonable to ask about the suitability of using the community home range as opposed to those smaller core areas in defining female chimpanzee space use. In fact, studies identifying core area space use recognized but excluded movements of females when they were with others or were cycling, both of which are associated with wider ranging within the community home range (Wrangham 1979; Hasegawa 1990; Williams et al. 2002; Emery Thompson et al. 2007). Moreover, the community home range appears to be important to female reproduction in that larger ranges are associated with shorter interbirth intervals (Williams et al. 2002). Finally, female western chimpanzees regularly range widely within the community home range (Lehmann and Boesch 2005). These facts suggest that the community home range is the area most relevant for our purposes. In both eastern and western chimpanzee populations, males initiate most boundary patrols and engage in aggressive intercommunity encounters (Boesch and Boesch-Acherman 2000; Boesch et al. 2008). Although female western chimpanzees are sometimes active in intercommunity aggression, community home range size is better determined by the number of males than overall group size (Boesch and Boesch-Acherman 2000). All the evidence suggests that males set the community home range boundaries.

The convergence of spider monkey social organization with that of chimpanzees is well documented (Fedigan and Baxter 1984; Wrangham 1987; Symington 1988; Chapman et al. 1995). In conventional terms, they are both considered to live in fission-fusion societies, although the utility of this term has been questioned recently in recognition of the range of expression in fission-fusion dynamics across species, even those living in cohesive groups, with the recommendation that the term be abandoned (Aureli et al. 2008). Nonetheless, the Variable Home Range Sharing classification system also picks up similarities between chimpanzee and spider monkey social organizations to the exclusion of other types of social organizations. Like male chimpanzees, male spider monkeys form coalitions to patrol and defend the boundaries of the home range, whereas female spider monkeys generally range less widely and may avoid the boundaries if they have infants (Symington 1988; Chapman 1990; Shimooka 2005; Aureli et al. 2006; Wallace 2008). Males have also been observed to behave aggressively toward females from other communities during incursions into the home ranges of those communities (Aureli et al. 2006). Like female chimpanzees, female spider monkeys do not appear to control their home range boundaries.

Wrangham (2000) came close to the Variable Home Range Sharing model when he argued that the extent to which female chimpanzees and bonobos move directly between fruit patches or feed between patches can explain their different degrees of gregariousness within communities, with bonobos foraging as they travel between fruit patches more than do chimpanzees. Although he did not use the same terminology, in essence, he argued that bonobos engage more than chimpanzees in opportunistic searching. He argued for a different effect, however. Instead of making home range expansion easier, as I argue, opportunistic searching would prevent faster individuals (typically males) from reaching fruit patches first, thereby reducing scramble competition and enabling females to travel more often with others. If this is true, then we might expect female chimpanzees in Taï National Park also to engage in more feeding between fruit trees, i.e., more opportunistic searching, than those in eastern Africa, since, like bonobos, Taï females are more gregarious than eastern chimpanzees (Boesch and Boesch-Acherman 2000). As far as I know, this has not yet been tested.

Although the Variable Home Range Sharing model does not predict female movement strategies of Bullied females because home range expansion is not determined by females, females still must have movement strategies. I am intrigued by Wrangham's description of the ways in which female chimpanzees and bonobos differ in their movements in that chimpanzees follow the directed travel movement strategy more than bonobos do. If you will recall from chapter 5, spider monkey movements fit Lévy walks better than Brownian walks, which suggests that they also engage more in directed travel than opportunistic searching.

Finally, because chimpanzees and spider monkeys are classic fission-fusion species, it is worth discussing if having a fission-fusion social structure somehow predisposes primates to be Bullied females. In hamadryas baboons, the "original" fission-fusion species (Kummer 1968), bands are the ecologically relevant unit for females because the one-male units and clans that make up bands share the same home range (Swedell et al. 2011). More females remain in their natal band than leave it, and thus, they might be considered Promoting females (Swedell et al. 2011; Städele et al. 2015). If so, then they should engage more in opportunistic searching than directed travel, which seems to be supported from random walks analyses (Schreier and Grove 2010; chapter 5). Females do sometimes disperse to other bands, but notably, it is forced upon them by males (Pines and Swedell 2011; Swedell et al. 2011). This suggests that they might instead be considered Bullied females. If this is the case, the Variable Home Range Sharing model can make no predictions about their movement strategies. To be accurately classified, it will be necessary to determine which sex controls the band's home range boundaries. Alternatively, hamadryas baboons may be an example for which the Variable Home Range Sharing model's classification of females does not work well at the species, or even population, level but needs an individualized approach (Jack and Isbell 2009).

Male Behavior and Distracted Females: Multiple Scenarios

Distracted females are those who disperse from their natal groups without being evicted or coerced and travel with other unrelated reproductive females. Their space use is expected to be determined ultimately by factors other than food. Female ashy red colobus monkeys appear to be a good example. While their ranging patterns are not, of course, insensitive to the locations of their foods (Struhsaker 1975, 2010; Isbell 1983, 2012; Chapman and Chapman 1999), females in some populations appear to be limited less by food and more by predation, especially by chimpanzees, even to the extent that their populations have approached local extirpation (Stanford 1998; Teelen 2008; Struhsaker 2010; Watts and Amsler 2013). The red colobus population in Gombe is among those that have experienced a severe decline from chimpanzee predation (Stanford 1998). Immature red colobus are most vulnerable to chimpanzee predation there (Stanford et al. 1994), which naturally reduces the reproductive success of their mothers. Female red colobus typically disperse to other groups without aggression from those in their former or new groups (Struhsaker 2010). They tend to be attracted to groups with more adult males (Struhsaker 2000), or perhaps more specifically, to groups with more males willing to counterattack chimpanzees (Stanford 1998:154). Male ashy red colobus form coalitions to counterattack chimpanzees (Stanford 1995, 1998; Struhsaker 2010; chapter 6), so females may choose to live in home ranges with more males or with more males willing to protect them and their offspring from chimpanzee predation rather than in home ranges where food quality or abundance is greatest.

Female mountain gorillas may provide an example of a species "distracted" by infanticide. Female mountain gorillas may disperse multiple times in their lives, and they gravitate toward males (Harcourt and Stewart 2007). Preferred males appear to be those who can best protect the females' offspring from infanticide (Robbins et al. 2009a). Although mountain gorillas are also sensitive to food and prefer to range and feed in high-quality areas, home range size is apparently related more to male–male competition than female energetic needs (Watts 1998), and female reproductive success appears to be constrained more by infanticide than food (Watts 1994; Robbins et al. 2007; Caillaud et al. 2020). Infanticide occurs when females with infants encounter, in the absence of the group's resident male(s), other males as solitary males or members of other groups (Watts 1989). The population grew substantially as human interventions to protect the gorillas were established (Caillaud et al. 2014). However, as the number of groups and group density increased over time, intergroup encounters also increased, and the rate of infanticide increased, which contributed most to the decline in the annual population growth rate—from 5% to 2.4% (Caillaud et al. 2014, 2020). Food competition appears to have had little effect on the changing population trend because food availability did not decline appreciably (Grueter et al. 2013), home ranges were relatively stable (Caillaud et al. 2014), and interbirth intervals did not increase (Caillaud et al. 2020).

Another species that may fit under the classification of Distracted females is the northern muriqui. In this species, males remain in the natal group, whereas females disperse before puberty to other groups or as solitaries (Strier and Ives 2012; Strier et al. 2015). At Caratinga, the population has grown over time from two groups to four and has expanded into new areas, including the terrestrial niche (Tabacow et al.

2009). Interbirth intervals have become shorter, resulting in greater fertility among females, suggesting that food is not limiting them yet (Strier and Ives 2012; Strier et al. 2015). The costs of dispersal appear to be minimal as predation is uncommon and the little aggression that is directed toward immigrant females is short-lived (Printes and Strier 1999; Bianchi and Mendes 2007; Tabacow et al. 2009). Still, there is no secondary dispersal; once a female settles with a different group, she remains there for her lifetime (Strier et al. 2015). Dispersal by juvenile females only once would be an expected strategy for avoiding inbreeding (Marsh 1979). Theoretically, because females transfer only once in their lifetimes, they could eventually mate with their sons who have matured in their adopted group. They resolve this by largely avoiding such matings and preferring to mate with extragroup males. Recent immigrants, in contrast, avoid mating with extragroup males, some of whom could be male relatives (Strier 1997). Available evidence thus suggests that in this population, the risk of inbreeding may be more important than food in determining female space use.

Primate Social Organizations: Putting Males and Females Together

We began building our understanding of primate social organizations by classifying female social organizations based on the extent to which they share their home ranges with other females. Their ability to share their home ranges may occur as a result of the relative ease with which females can expand their space, and that may be dependent upon their movement strategies. The classification system is based on space use, an ecological characteristic that profoundly affects reproductive success for most female primates. Males largely mirror females in their movement strategies, and we can now overlay male space use onto female social organizations. The result of males and females moving through their space is what we recognize as primate social organizations.

Main Points

Males who are successful in obtaining mates must have already been successful in obtaining enough food to allow them to compete effectively against other males. In most cases, males and females share the same movement strategies, and this affects their social organization. Dispersers who mainly employ directed travel will benefit most by finding others who know about the locations of foods, whereas dispersers who mainly employ opportunistic searching are not as dependent on the knowledge of others to find their foods.

Consequently, Isolationist females and males are expected to pair up, but Generous females and males need not, and home ranges of such males will overlap those of multiple females. Constraining females frequently have more males than expected based on the numbers of females in their groups. Establishing and maintaining a home range seems critical for reproduction among Constraining females, but they appear to need male assistance to succeed. Numbers of males in Promoting female groups are not unusual in the context of sexual selection, with roughly one adult male per two females. In many species of Promoting females, females are actively engaged in aggressive intergroup encounters and so do not appear to benefit as much as Constraining females from male involvement in them.

Finally, there are females for whom movement strategies cannot be predicted because males control their home range boundaries by patrolling and aggression toward females (Bullied females) or because food takes a back seat to other selective pressures that affect their space use more (Distracted females). In the latter case, female home range selection appears to be more dependent on the locations of males who can protect them from predation or infanticide, or who are not close relatives. With the recognition that both competition for females and male movement strategies affect male space use and that sometimes males affect female space use, we have now overlayed male contributions onto female social organizations, allowing us to understand more fully the functional causes of diversity in primate social organizations.

CHAPTER 9

Answers to Some Questions about the Model

A new scientific truth does not triumph by convincing its opponents and making them see the light, but rather because its opponents eventually die and a new generation grows up that is familiar with it.

—M. Planck (1949:25)

The epigraph to this chapter is often paraphrased as, "Funeral by funeral, science marches on." It is always challenging for new ideas to make it into the mainstream of science, and primatology is no different in that way. The current views on explaining diversity in primate social organizations are, by now, so well ingrained in our classrooms, journal papers, and academic books that they are often treated as established facts rather than working hypotheses. Thus, we read statements such as the following: "All behavioral ecologists agree that primary benefits of living in social groups relate to mechanisms of predator detection and defense . . . and to enhanced discovery and defense of resources" (Fragaszy et al. 2004:51). Please do not let that discourage you from entertaining new ideas. I hope I have convinced you (especially younger) readers that the status quo is, in fact, unsatisfactory and that serious consideration of new ideas is warranted. As you have progressed through the chapters, you may have developed some questions about the alternatives I am proposing. The questions below are ones I asked myself as I developed the Variable Home Range Sharing model. Perhaps you will have others. Perhaps they will also lead you to find the fatal flaws or develop more tests to fine-tune the model.

Don't Movement Strategies Just Reflect Food Distributions?

The answer to this question is, in a nutshell, yes, but not in the way we currently measure food distributions (see below). Food distribution has long been considered important in determining group living and group size but also female social relationships within and between groups (Wrangham 1980; Isbell 1991; Sterck et al. 1997; Koenig 2002). Specifically, it has been argued that group living evolved as a coalitionary response to clumped foods such as fruits, with larger groups gaining a competitive advantage over smaller groups because they can more successfully acquire and maintain access to clumped food trees (Wrangham 1980). Food distribution has also been indirectly associated with female philopatry. The argument is that clumped foods favored the evolution and continued maintenance of female groups and that inclusive fitness benefits favored coalitions of kin in intergroup competition, whereas evenly distributed foods such as leaves are not worth competing for and would not have favored the evolution of female philopatry or coalitions of kin (Wrangham 1980; van Schaik 1989). From there, food distribution was also considered, along with predation, to be responsible for variation in female social relationships beyond cooperative coalitions, including the expression of "competitive regimes" within and between groups (Sterck et al. 1997).

I reiterate that I am not focused on the details of female social relationships. The Variable Home Range Sharing model only covers the basics in socioecology, i.e., to what extent do females share home ranges, whether females should group or not, how large should groups be, and what accounts for variation in male contributions to female social organizations. In addition, the proposed selective pressure for grouping (sunlight allowing visual monitoring in shared home ranges) would have favored grouping ahead of the evolution of female social relationships in and between groups. Once females were brought together by visual monitoring, social selective pressures could then begin to operate. This is consistent with the argument that ecological conditions alone are unlikely to explain nuances in female social relationships across species (Koenig 2002; Thierry 2008, 2021).

Given the long theoretical importance of food distribution in primate socioecology however, a great deal of effort has been devoted to measuring it in a variety of ways (reviewed in Vogel and Dominy 2011). Because the habitats in which most primates live are highly complex, primatologists have tended to focus on characterizing food distribution only in relation to "important" foods, determined by the prominence of those foods in the diet (e.g., Koenig et al. 1998). They are often limited to one food type, e.g., fruits or leaves, because primates tend to eat one type more than another (e.g., Vogel and Janson 2011). These broad dietary categories might help us guess at movement strategies because fruits, leaves, and arthropods tend to be distributed differently, but we cannot rely on them to accurately predict movement strategies. For example, a "frugivorous" primate might be expected to move more slowly and opportunistically when searching for insects compared to fruits, but woolly monkeys, *Lagothrix lagotricha*, actually travel farther and faster when foraging for insects than when they feed on fruits (Di Fiore 2003). Recall also that, within broad food categories, foods may be distributed differently for different species, e.g., in Kibale, leaves are clumped for black-and-white colobus and more widely distributed for red colobus (Struhsaker and Oates 1975; chapter 1). Nor does the designation of "frugivore" tell us if such a female will employ directed travel or opportunistic searching between "important" fruit trees. When minor foods, or foods that are discovered as individuals make their way to "important" foods, are not considered, it will result in an incomplete picture of food distribution. Movement strategies get around this issue by reflecting the distribution of all foods eaten, not just the important foods. An alternative method of determining total food distribution using the movements of individuals has already been advocated (Isbell et al. 1998b).

Most importantly, broad dietary categories and their associated distributions of foods as currently measured do not tell us about the ability of females to expand and share their home range with other reproductive females. Found among Isolationists, for example, are frugivores (e.g., titi monkeys and siamangs: Van Belle et al. 2016, 2020; Lappan et al. 2021), folivores (e.g., indris: Bonadonna et al. 2017, 2020), gummivores (e.g., fork-marked lemurs: Schülke and Kappeler 2003, Schülke 2005), and faunivores (e.g., spectral tarsiers: Gursky 2010). It would be very surprising to find that their important foods are distributed similarly for all those species. I think it would be less surprising to find that all of them require certainty in the locations of their foods to maintain their foraging efficiency and that they use directed travel to get to those foods.

How Does the Variable Home Range Sharing Model Differ from Other Foraging Efficiency Models for Group Living?

Cody (1971) and Altmann (1974) proposed that group living in birds and primates, respectively, could arise by individuals staying with others to minimize returns to depleted food patches. This is close to what I am proposing, but instead of minimizing returns to patches already fed in, I am suggesting visual monitoring facilitates grouping as a way to minimize foraging along paths others are taking. Ward and Zahavi (1973) proposed that widely and patchily distributed foods favor foraging and roosting flocks of birds as a way for individuals to obtain and share information about the locations of foods. Single birds increase their foraging efficiency by seeing a flock already feeding and then joining it or by following others from the roost to a food patch they themselves have not yet discovered. This argument does not explain cohesive groups with stable membership, however. Among primates, information sharing about food seems to be limited (Alexander 1974; chapter 1) but may best be found in species whose members share the same home range but do not live in cohesive groups, such as chimpanzees, whose pant-hoot vocalizations draw others to rich food sites (Wrangham 1977; Notman and Rendall 2005). All three models require foods to be patchily distributed, but with the Variable Home Range Sharing model, foods can be distributed in any way for females sharing a home range. Furthermore, those models do not explicitly identify sunlight as necessary for the formation of groups (although this can be inferred).

Does Phylogeny Factor Into
the Variable Home Range Sharing Model?

With the Variable Home Range Sharing model, phylogeny is important in that related taxa are likely to be more similar in physiological
requirements, morphological characteristics, and diets than unrelated
taxa (Wilson 1975), which will affect how they move through their environments. Related taxa are thus likely to be similar in the extent to
which they share their home ranges and, ultimately, in their social
organizations. For instance, from available evidence, cercopithecines are
Promoting females, callitrichines are Intolerant females, and howler
monkeys are Constraining females. However, Constraining females (as
well as Isolationists and Generous females) also include unrelated taxa,
which reveals behavioral convergence independent of phylogeny and the
ability of ecological factors to transcend phylogeny in determining
movement strategies and home range sharing.

Several models have been proposed to explain the evolutionary trajectory of primate social organizations. The traditional one proposes
that, from the basal mammalian social organization of a solitarily
foraging mother with offspring, selection favored increased complexity
leading to large and cohesive multi-female groups (Crook and Gartlan
1966; Eisenberg et al. 1972; Terborgh and Janson 1986; Rodman 1999).
Another one proposes that primates initially split into group living and
solitary foraging because of a difference in activity periods and strategies
for avoiding predators. Diurnal primates employed early detection and
escape, which led to living in groups, and nocturnal primates employed
concealment, which led to living solitarily. Following that initial split,
pair living evolved where predation pressure is low (van Schaik and van
Hooff 1983). A more recent one proposes that the evolutionary path went
instead from solitary foraging directly to large multi-male, multi-female
groups, from which pair living or single-male, multi-female groups
branched off, all driven by phylogenetic constraints (Shultz et al. 2011).

Although the Variable Home Range Sharing model assumes the an-
cestral primate arose from a mammal that was a nocturnal solitary forager, it does not envision a stepwise series of changes that led from one

to another of the various types of female social organizations. However, it does envision two basic alternatives—to share or not to share the home range—and the alternative taken would have depended on how much females need food location certainty and thus how much they engage in directed travel versus opportunistic searching movement strategies. The non-sharing pathway employs more directed travel and is represented by Isolationists, who do not share their home range at all, Intolerant females, who can share their home range entirely with other adult females but none that are reproductively active, and perhaps Constraining females, who share their home range entirely but with limited numbers of reproductive females. These three categories are expected to show a graded expression of directed travel, with the heaviest reliance among Isolationists, and the least, among Constraining females.

The sharing pathway employs opportunistic searching more than directed travel and is represented by Promoting females, who share their home range entirely with many other females, and perhaps Generous females, who may share their home range partially with many other adult, reproductive females. (Since no predictions can be made about the movement strategies of Bullied and Distracted females, they could belong to either pathway.) Disregarding Bullied females because males control home range size and also likely affect female movements (Williams et al. 2002), females that fully share their home range are always diurnal, but those that share their home range only partially with other females are nocturnal with one exception (orangutans; chapter 7). Being diurnal means it is possible to use visual monitoring to avoid following paths already being taken as other females feed, allowing them to maintain their foraging efficiency even while sharing their home range completely with other females. It is likely that ecological selective pressures operating on body size (chapter 7) influence the diurnal–nocturnal and Promoting–Generous female splits more so than phylogeny.

It might have been a relatively simple process to shift from being a Generous female to a Promoting female as body size increased and diurnality became an available niche. Perhaps there was an evolutionary trend with both pathways to increase opportunistic searching because that movement strategy appears to be more flexible and forgiving than directed travel. Given the widespread expression of some of the female categories across primates, it seems unwarranted to suggest more than this.

Aren't "Constraining" and "Promoting" Females Just Different Names for Specialists and Generalists?

If fieldworkers conducting a survey of a new species of primate find females living in cohesive groups, they would not necessarily be able to identify the females as Constraining or Promoting; the only difference between them on the surface is in group size—Constraining females tend to have smaller maximum group sizes than Promoting females. Deeper down, however, according to the Variable Home Range Sharing model, this difference in maximum group size is determined by differences in the ways that they move. Constraining females are predicted to employ more directed travel than Promoting females, which will limit their ability to expand their home ranges into new areas to accommodate other reproductive females.

Generalists have a broad diet and can shift their food choices when food availability changes, whereas specialists are restricted to a narrower range of foods (Takahashi et al. 2019). Baboons, macaques, and mangabeys are considered generalist omnivores (Homewood 1978). Homewood (1978:376) described them as spending "considerable time searching both visually and by object manipulation" whereas specialist folivores show "minimal search and handling time." Her description of Tana River mangabeys, *Cercocebus galeritus*, as generalists based on their locomotion, frequent manipulation of potential foods, broad use of structural supports, and diversity of food species, is consistent with the opportunistic searching movement strategy of Promoting females. Indeed, we can imagine that generalists might employ opportunistic searching more than specialists because it would be more challenging to obtain a wide variety of foods with directed travel.

However, specialists are not limited to Constraining females (many of which are folivores), and generalists are not limited to Promoting females. For instance, Isolationist gibbons are often called ripe-fruit specialists (Clink et al. 2017), and Intolerant pygmy marmosets are gum specialists (Power 2010). Gray mouse lemurs are Generous females and also dietary generalists (Dammhahn and Kappeler 2008; but see Radespiel et al. 2006). The generalist–specialist dichotomy is thus not synonymous only with Constraining and Promoting females.

In addition, although the generalist–specialist concept and movement strategies are likely related, the generalist–specialist concept reveals nothing about whether generalists are more or less likely than specialists to share their home ranges with other reproductive females. If there is a dichotomy between generalists and specialists in relation to female social organizations, it might split along the lines of Isolationists, Intolerant females, and Constraining females tending to be specialists, and Generous and Promoting females tending to be generalists, according to their predominant movement strategies, although describing the divide as such would be limiting the reach of the Variable Home Range Sharing model.

Aren't Physical Constraints of the Habitat Responsible for Differences in Movement Strategies?

One criticism that might be leveled against a focus on movement strategies as a determinant of the extent to which females can share their home ranges is that female movements may be dictated more by the physical characteristics of their environment (Cannon and Leighton 1994; Di Fiore and Suarez 2007; Hopkins 2011; McLean et al. 2016; de Guinea et al. 2019; Abreu et al. 2021) than by the relative importance of knowing where foods are located in time and space. For example, smaller primates may be able to use a wider variety of supports than larger primates (Fleagle and Mittermeier 1980; McGraw 1998), or different strata of arboreal environments may favor different ways of traveling (Fleagle and Mittermeier 1980).

Using LiDAR (light detection and ranging) to characterize the forest canopy on Barro Colorado Island, Panama, and GPS to track the movements of mantled howlers, white-faced capuchins, and Central American spider monkeys, researchers confirmed behavioral observations that capuchins and spider monkeys move faster and farther per day than howler monkeys (McLean et al. 2016). They also found that all species avoided gaps in the canopy, that capuchins and spider monkeys preferred taller canopies, and that howler monkeys and spider monkeys preferred thicker crown densities (McLean et al. 2016). Clearly, these sympatric species are using the same forest in different ways, but it is unclear how—if at all—these differences affect the ability to employ directed travel or opportunistic searching.

It is probably impossible to answer the question of whether the physical environment affects movement strategies of different species, even when they are sympatric, because different species are likely to use different supports as they move along. Possibly the only way to determine if movement strategies are caused by structural constraints would be to conduct controlled experiments in which the supports are identical for individuals from different social organizations. If they still move in different ways, then we can rule out structural constraints as the cause of those differences. Fortunately, this has been done. In a series of elegant experiments with primates in a large enclosure, Dorothy Fragaszy (1980) systematically controlled arboreal supports and routes, making them equally available to coppery titi monkeys, *Plecturocebus cupreus* (formerly *Callicebus moloch*), and black-capped squirrel monkeys, *Saimiri boliviensis* (formerly *S. sciureus* or *S. cassiquiarensis*; D. Fragaszy, pers. comm.), and then challenged them to move along the supports to reach a goal. Titi monkeys were more conservative in taking the routes, and they used certain pathways habitually even when shorter routes to the goal were available, whereas squirrel monkeys more often took novel, shorter pathways to the goal. These species differences are consistent with the categorization of female titi monkeys as Isolationists and of black-capped squirrel monkeys as Promoting. Isolationists are predicted to employ more directed travel than opportunistic searching, leading to difficulty in expanding their home range into new areas, and sticking to familiar routes would make home range expansion even more difficult. Promoting females, on the other hand, are predicted to employ more opportunistic searching than directed travel, leading to greater ease in the ability to expand their home range into new areas, and taking novel routes would facilitate home range expansion. Indeed, Fragaszy (1980:31–32) essentially stated as much in her assessment of how the differences she found between titi monkeys and squirrel monkeys relate to their natural environments: "maintenance of a small territory, as seen in titis, requires strong attraction to place in favor of interest in novel places. . . . Squirrel monkeys, on the other hand, are nomadic and much more active than titis. Attraction to novel places and pathways and disregard for minor increases in travel distance are consistent with these habits. In the long run, squirrel monkeys' apparently more random travel patterns may result in efficient use of unfamiliar resources."

How Can We Test the Variable Home Range Sharing Model Further?

First, we need data on the extent of home range sharing among females for more species to be able to classify all primates. Adult females living by themselves in their own home ranges, with or without home range overlap with other females, will be easy to classify as Generous and Isolationist females, respectively. For group-living females, however, there is the risk of misinterpretation. It is very important to note that I am not referring to home range overlap *between* groups but to home range sharing between females *in the same* group. I know it sounds strange to separate home range sharing from group living because of course females living in the same group share the same home range, but it might not have been that way with the first diurnal primates. Fortunately, the extent of home range sharing is basic information for field studies.

Second, we need data on who is responsible for the size of the home range to distinguish Bullied females from other categories. Female control would be suggested if home range size is sensitive to the number of females in the group, if females are regularly engaged in aggressive intergroup encounters, or if females reward males for their participation in aggressive intergroup encounters. Male control would be suggested by boundary patrols and aggression toward females that restricts female movements within their home range.

Third, we need data on the number of adult females that reproduce in a given home range. This is important for distinguishing Intolerant females from the other categories of group-living females. If only one typically reproduces even though there is often more than one adult female in the home range, then they can be classified as Intolerant regardless of the overall group size.

Fourth, we need data on the process of female dispersal and its consequences. Do females disperse from their natal groups, and if they do, is it voluntary or are they evicted? For example, female woolly monkeys are known to disperse from their natal groups (Di Fiore and Fleischer 2005; Di Fiore et al. 2009), but are they forced out, as Stevenson and colleagues (2015) have suggested, or do they leave on their own? This information will help us separate Constraining females from Distracted females.

Finally, for species that are found both in pairs and small multi-female groups, we need to know more not only about the process of female dispersal, but also about the ability of adult females to expand the home range to accommodate other adult females. Observations of only one adult female per home range would suggest they are Isolationists, but how do we classify them when two or three adult, reproductive females share the same home range? White-faced sakis, *Pithecia pithecia,* are a prime example of this variation (Lehman et al. 2001; Norconk 2006; Norconk 2011; Thompson 2016). Females disperse, as indicated by the presence of floaters (Thompson and Norconk 2011), but females have also been observed to remain and reproduce in their natal groups (Norconk 2006). Do their home ranges expand with additional adult, reproductive females, as seen in golden lion tamarins and black-crested gibbons with more than one reproductive female (Charles-Dominique 1977; Hankerson and Dietz 2014; Guan et al. 2018)? If so, they might be Constraining females rather than Isolationists. If this is the case, confirmation would be possible if targeted aggression is observed to precede natal dispersal at some point as female numbers in a group increase, as occurs in red howlers (Crockett 1984; Pope 2000). On the other hand, if home range size does not increase with additional adults (typically daughters) such that these "extra" individuals negatively affect the mother's reproduction, as occurs in fork-marked lemurs (Schülke 2003), they should probably remain classified as Isolationists.

While we are classifying females into their various social organizations, we also need to collect the data to test predictions of the Variable Home

Range Sharing model. Most importantly, we need to quantify move-ment strategies. There may be several ways in which to do this, includ-ing those mentioned in chapter 5—random walks analyses and activity budgets, the latter either as percentages of time females spend traveling and foraging or as bout lengths of locomotion during travel and feed-ing. I previously suggested that collecting data on "fast travel" and "slow foraging" as Charlie Janson and Mario Di Bitetti (1997) have done, would also be useful (Isbell 2004). All the above would be easier than my other suggestion to record successful and unsuccessful stops for food, with the prediction being that successful stops would be more common with directed travel (then called goal-directed travel) than opportunis-tic searching (then called wandering) (Isbell 2004). I no longer recom-mend this approach because my students found collecting the data too difficult in arboreal habitats with poor visibility. With more data available on movement strategies, including operational definitions of traveling and foraging that distinguish between directed travel and op-portunistic searching, and activity budgets that separate the sexes, we will be able to test more fully the prediction that the different female social organizations are associated with predictably different proportions of directed travel and opportunistic searching.

It would also be exceedingly helpful if more controlled experiments on comparative movement strategies and support use could be carried out along the lines of Fragaszy's (1980) work on titi and squirrel monkeys.

How Might the Variable Home Range Sharing Model Inform Primate Conservation?

Determining the type of female social organization for a given species might have real life benefits for primate conservation. Since the contin-ued survival of primates requires sufficient land, we need to understand more fully the constraints and affordances different female social organizations place on land use. For example, one might predict that population sizes of Promoting females will increase when food abun-dance increases but that population sizes of Distracted females will in-crease even while food abundance remains the same if they are somehow protected from the non-food factor that is hypothesized to keep them in check, e.g., predation or infanticide (until food eventually does limit them). As predicted for Promoting females, a population of white-faced capuchins at Santa Rosa National Park, Costa Rica, grew as it expanded into regenerating areas, which presumably increased their food abundance (Fedigan and Jack 2001). Long-term studies of Distracted females have also reported population increases for mountain gorillas (Caillaud et al. 2014, 2020); northern muriquis (Dias and Strier 2003; Tabacow et al. 2009; Strier and Ives 2012); and ashy red colobus (Struhsaker 1975; Snaith and Chapman 2008; Snaith et al. 2008; Isbell 2012; Gogarten et al. 2014, 2015; but see Chapman et al. 2015). Unfortunately for our purposes here, the availability of their foods also increased, which either means these populations do not support the prediction, or they do not meet the criterion for testing it. Conducting long-term studies of Distracted females in isolated and fully occupied forests where dispersal is limited and their predators have recently been extirpated might be one, albeit challenging, way to test this prediction.

Another prediction is that groups of Constraining females will become more numerous as targeted females disperse and attempt to form their own groups when populations grow, whereas groups of Promoting females will become larger more often rather than more numerous. Studies of ring-tailed lemurs and mantled howlers (Jolly and Pride 1991; Milton 1980; Fedigan and Jack 2001)—both categorized as Constraining—and white-faced capuchins, gray-cheeked mangabeys, and blue monkeys (Olupot et al. 1994; Fedigan and Jack 2001; Angedakin and Lwanga 2011)—all categorized as Promoting—support this prediction, but more data are needed. Unfortunately, this prediction will also be difficult to test because, in addition to requiring long-term monitoring, most primate populations are declining now (Estrada et al. 2017).

Referring back to what I wrote in the Preface, more fully understanding socioecology, with its focus on interactions between animals and their environments, should help us predict future interactions in ways that could reduce the effects on them of our own human population growth. With so many primate species now threatened with extinction

(Estrada et al. 2017), we likely cannot save them all. It is important for us, therefore, to predict which ones are most in need of our intervention and which of those are most likely to survive if we do intervene. Knowledge of female social organizations based on the Variable Home Range Sharing model may help answer the question of *why* species *naturally* vary in their susceptibility, all else being equal. For example, we might expect Isolationists and Intolerant females to fare more poorly than Promoting females during translocations because of their heavier reliance on food certainty.

Moreover, if we choose to put our efforts into protecting populations of Isolationists and Intolerant females *in situ*, then we may need to recognize that they will be more challenging to protect than Promoting females if they are forced to move away from deteriorating home ranges. Alejandro Estrada and colleagues (2017) highlighted the ability of some genera, e.g., *Macaca*, *Papio*, and *Semnopithecus*, to adjust easily to changing environments. It is notable that these genera typically consist of Promoting females. Their success in adapting may be related, in part, to their not needing food location certainty, as suggested by their extensive use of the opportunistic searching movement strategy (which accommodates food location uncertainty), thus making it easier to survive in new areas.

Final Points

I hope I have provided the rationale and enough evidence to challenge current classifications of primate social organizations and the ecological explanations for them. I hope other primatologists will see merit in the functionality of the Variable Home Range Sharing classification system of primate social organizations and will adopt it regardless of whether differences in movement strategies, the proposed mechanism for variation in home range sharing, hold up. I also hope that movement strategies will be taken seriously enough that the data will be collected, and the predictions tested, at least by younger primatologists. I have offered a few suggestions for such tests, and I hope to see new, more creative, tests in the future. Good luck and have fun!

Activity Budgets from a Literature Search
That Were Reviewed to Test the Hypothesis
That Relative Time Spent in Directed Travel
Versus Opportunistic Searching
Can Predict Female Social Organizations

Scientific Name[1]	Common Name	Study Site	Travel (%)	Forage (%)	Feed (%)
Alouatta caraya	Black-and-gold howler monkey	El Pinalito Provincial Park, Argentina	15.0		15.0
Alouatta caraya	Black howler monkey	Estancia Casa Branca, Brazil	17.0		17.0
Alouatta guariba (*A. clamitans*)	Brown howler monkey	RPPN Feliciano Miguel Abdala, Brazil	15.0		18.5
Alouatta guariba	Brown howler monkey	El Pinalito Provincial Park, Argentina	11.0		14.0
Alouatta palliata	Mantled howler	Barro Colorado, Panama	10.2		16.2
Alouatta palliata	Mantled howler	Los Tuxtlas, Mexico	1.0		17.0
Alouatta palliata	Mantled howler monkey	La Suerte, Costa Rica	11.3		12.5
Alouatta palliata (*A. villosa*)	Mantled howler monkey	Barro Colorado, Panama	9.2		10.0
Alouatta pigra	Central American black howler monkey	Monkey River, Belize	7.5		18.6
Alouatta pigra	Central American black howler monkey	Campeche, Mexico	9.8–13.1		14.2–20.8
Alouatta seniculus	Red howler monkey	Yotoco Reserve, Colombia	10.4		22.6
Alouatta seniculus	Red howler monkey	Finca Merenberg, Colombia	5.6		12.7
Alouatta seniculus	Red howler monkey	Tinigua NP,[2] Colombia	15.0		23.0
Aotus azarae	Azara's owl monkey	Beni Biological Reserve, Bolivia	18.9		31.7

Activity categories and relevant notes	Source
See Table 5.3.	Agostini et al. 2012
Locomotion, feeding, social drinking, resting; not defined. Percentages for adult females only; estimated from graph.	Bicca-Marques and Calegaro-Marques 1994
Moving (within the same tree),[3] traveling (between trees), feeding (inspection of food, bringing it to the mouth, chewing, and swallowing). Percentages averaged from two groups.	Jung et al. 2015
Moving (changing spatial position, only including short, nondirectional movements in cases in which the group is engaged in nontraveling activities), traveling (changing location directionally, in a context of group traveling in a goal-oriented direction), feeding (procuring, handling, ingesting, or chewing any food item), resting, social, other. Percentages estimated from graph of maxima for two groups.	Agostini et al. 2012
Travel, feed, rest; not defined.	Milton 1980
Travel (synchronous mobilization of the troop away from the tree or group of trees to another area within the forest fragment), locomotion (not defined), feeding (not defined), resting, social interactions.	Estrada et al. 1999
Travel (move along a substrate not while feeding), feed (manipulate food or water with hands, feet, or mouth), rest, social, other.	Schreier et al. 2021
Movement (walk, run, swing and grasp, leap, brachiate, bipedal walking, hanging under branches, shifting along branches), feeding (food in mouth or in hand; feeding while moving was feeding), resting, vocalization, urination, defecation.	Richard 1970
Locomote (not defined), feed (not defined), inactive, social.	Pavelka and Knopff 2004
Feed, rest, move, and social; not defined. Data from adult females in lactating and non-lactating conditions.	Dias et al. 2011
Moving (any time the focal individual was traveling without feeding), feeding (not defined). Percentages averaged from two groups.	Palma et al. 2011
Move (walk, run, climb, jump), feed (while lying prone, sitting, standing, hanging by tail), rest, carry, groom, eliminate, copulate, nurse, play. Percentages averaged from two groups.	Gaulin and Gaulin 1982
Moving, feeding, resting, social; not defined.	Stevenson et al. 2000
Locomotion (walk from one branch to another), feeding (also including short walks in search of food), resting.	Garcia and Braza 1987

Scientific Name[1]	Common Name	Study Site	Travel (%)	Forage (%)	Feed (%)
Aotus lemurinus	Colombian owl monkey	Manizales, Colombia	38.4		
Aotus miconax	Peruvian owl monkey	Comunidad Campesina Yambras-bamba, Amazonas Department, Peru	54.0		33.0
Ateles belzebuth	White-bellied spider monkey	Maracá Ecological Station, Roraima, Brazil	36.0		18.0
Ateles belzebuth	White-bellied spider monkey	Tinigua NP, Colombia	27.0		25.0
Ateles belzebuth	White-bellied spider monkey	Yasuní NP, Ecuador	24.9		16.7
Ateles chamek (*A. belzebuth*)	Black spider monkey	Manu NP, Peru	26.0		29.0
Ateles chamek	Black spider monkey	Noel Kempff Mercado NP, Bolivia	29.7		18.9
Ateles geoffroyi	Central American spider monkey	Runaway Creek Nature Reserve, Belize	25.3		24.2
Ateles geoffroyi	Central American spider monkey	Marques de Comillas and Montes Azules Biosphere Reserve, Mexico	12.0		44.0
Ateles geoffroyi	Central American spider monkey	Barro Colorado, Panama	27.6		10.8
Ateles geoffroyi	Central American spider monkey	Santa Rosa NP, Costa Rica	32.6		33.5
Avahi laniger	Eastern woolly lemur	Ranomafana, Madagascar	13.5		22.0
Avahi meridionalis	Southern woolly lemur	Sainte Luce, Madagascar	14.0		15.0
Brachyteles arachnoides	Southern muriqui	Barreiro Rico, São Paulo, Brazil	9.7		
Brachyteles arachnoides	Southern muriqui	Parque Estadual Carlos Botelho, São Paulo, Brazil	22.2		28.3

Activity categories and relevant notes	Source
Traveling (change positions [moved] while foraging or engaged in another behavior either within or between trees), feeding and foraging (movements associated with the search and consumption of food), social, rest, other. Percentages averaged from two groups. Combined foraging and feeding: 28.6%.	Bustamante-Manrique et al. 2021
Traveling, feeding, resting, other; not defined.	Shanee et al. 2013
Traveling (within and between crowns), feeding (including time spent searching for and manipulating food items), resting.	Nunes 1995
Moving, feeding, resting, social; not defined.	Stevenson et al. 2000
Traveling, foraging, resting, socializing; not defined	Suarez 2006
Traveling (not defined), feeding (including all movements and positional changes within a food tree), resting, other.	Symington 1988
Traveling, feeding, resting, other; not defined.	Wallace 2001
Travel, feed, inactive, social, other; not defined.	Hartwell et al. 2014
Traveling (movement between tree crowns or within the crown of a tree that was not directly food related), feeding (masticating or consuming food items), resting, other.	Chaves et al. 2011
Movement (walk, run, swing and grasp, leap, brachiate, bipedal walking, hanging under branches, shifting along branches), feeding (food in mouth or in hand; feeding while moving was feeding), resting, vocalization, urination, defecation.	Richard 1970
Traveling, resting, feeding; not defined.	Chapman et al. 1989
Traveling, feeding, resting, grooming; not defined.	Harcourt 1991
Moving, feeding, resting, other; not defined. Percentages averaged from four individuals.	Norscia et al. 2012
Traveling (not defined), feeding (included inspection of food), resting, other. Combined foraging and feeding: 27.8%.	Milton 1984
Traveling (any mobile activity, whether walking, climbing, running, or leaping), feeding (the manipulation and consumption of any plant material), resting, socializing.	Talebi and Lee 2010

Scientific Name[1]	Common Name	Study Site	Travel (%)	Forage (%)	Feed (%)
Brachyteles hypoxanthus (B. arachnoides)	Northern muriqui	Fazenda Montes Claros, Minas Gerais, Brazil	29.4		18.8
Brachyteles hypoxanthus (B. arachnoides)	Northern muriqui	Estação Biológica de Caratinga, Minas Gerais, Brazil	17.5		40.0
Brachyteles hypoxanthus	Northern muriqui	Reserva Particular do Patrimônio Natural Feliciano Miguel Abdala (Estação Biológica de Caratinga), Fazenda Montes Claros, Minas Gerais, Brazil	26.2		21.5
Callicebus coimbrai	Coimbra-Filho's titi monkey	Fazenda Trapsa, Sergipe, Brazil; Mata do Junco Wildlife Refuge, Sergipe, Brazil	134–155 min	19–95 min	93–168 min
Callicebus nigrifrons	Black-fronted titi monkey	Serra do Japi Municipal Ecological Reserve, Jundiaí, Brazil	24.0		35.0
Callimico goeldii	Goeldi's monkey	Tahuamanu Biological Field Station (San Sebastian), Bolivia	17.0 and 32.0	6.0 and 4.0	9.0 and 8.0
Callithrix flaviceps	Buffy-headed marmoset	Feliciano Miguel Abdalla Private Natural Heritage Reserve (Caratinga Biological Station), Brazil	26.9	25.4	13.7
Callithrix flaviceps	Buffy-headed marmoset	Augusto Ruschi Biological Reserve, Espírito Santo, Brazil	39.1	37.91	6.1
Callithrix geoffroyi	Geoffroy's tufted-ear marmoset	Aracruz Celulose S.A., Espírito Santo, Brazil	20.4	13.4	21.3

Activity categories and relevant notes	Source
Travel (any movement involving a change in location, both within and between trees), feeding (procuring and handling food items with the hands or mouth; chewing or obvious ingestion), rest, socializing, vocalizing, other.	Strier 1987
Traveling (both within and between tree canopies), feeding (ingestion of fruits, flowers and flower products, leaves, tree bark, bamboo, ferns, and unidentified items), drinking water from ground sources or holes in tree trunks, resting, socializing. Percentages from adolescent females only.	Printes and Strier 1999
Travel, feed, rest, social, drink; not defined. Females only, including juveniles.	Tabacow et al. 2009
Move (mobile activity: walking, climbing, running, leaping), forage (localized searches for plant or animal material), feed (mastication and consumption of plant and/or animal material), rest, social. Values are mean minimum and maximum minutes from two groups in a large and small forest fragment.	Souza-Alves et al. 2021a
Traveling (not defined), feeding (seeking food, which includes close inspection of food resources, and subsequent manipulation, processing, chewing, and swallowing), resting, other.	Caselli and Setz 2011
Travel (quadrupedal movement including walking and running, branch-to-branch leaping, vertical clinging and leaping trunk to trunk, and vertical ascent and descent), forage (searching for a food object within a substrate, and visual search of the ground or plant for a food object), feed (eating or manipulating a food item), rest, other. Percentages from two studies at same study site several years apart.	Porter and Garber 2004; Porter et al. 2007
Move, forage, feed, rest, other; not defined. Percentages based on scan sample records.	Ferrari and Hilário 2014
Move, forage, feed, rest, other; not defined. Percentages based on scan sample records.	Ferrari and Hilário 2014
Moving (any mobile activity, whether walking, climbing, running, or jumping), foraging (any activities involved in searching for animal food, including visualizing or manipulating), feeding (the consumption of plant and animal matter), gouging (gouging holes in the bark and branches of trees but not eating gum), resting, other.	Passamani 1998

Scientific Name[1]	Common Name	Study Site	Travel (%)	Forage (%)	Feed (%)
Callithrix jacchus	Common marmoset	Mata do Junco Wildlife Refuge, Sergipe, Brazil	23.7	40.9	9.8
Callithrix jacchus	Common marmoset	Baracuhy Biological Field Station, Paraíba, Brazil	7.0	35.0	11.0
Callithrix kuhli	Black tufted-ear marmoset	Lemos Maia Experimental Station, Cocoa Research Station, Bahia, Brazil	38.0	25.0	23.0
Cebuella pygmaea	Pygmy marmoset	San Pablo; Sacha; Amazoonico; Zancudococha, Ecuador	18.0—20.0	5.0—12.0	20.0, 23.0, 33.0, 34.0
Cebus aequatorialis (*C. albifrons*)	Ecuadorian white-fronted capuchin	Tiputini Biodiversity Station, Yasuní NP, Ecuador	25.0	54.0	10.0
Cebus albifrons	Humboldt's white-fronted capuchin	Manu NP, Peru	21.0	39.0	22.0
Cebus imitator (*C. capucinus*)	Central American white-faced capuchin	Santa Rosa NP, Costa Rica	17.0	53.0	
Cebus kaapori	Ka'apor capuchin	Goianésia do Pará, Pará, Brazil	47.4	15.3	24.0
Cebus olivaceus	Wedge-capped capuchin	Funco Pecquario Masaguaral, Venezuela	20.0		

Activity categories and relevant notes	Source
Move (vertical or horizontal locomotion, including jumps between branches and trees, not accompanied by any other activity), forage (searching for food through visual scanning in a saltatory manner or manually searching the substrate, such as bark on tree trunks or leaves on the ground), feed (chewing on or ingesting animal or plant item), rest, social. Percentages calculated from total minutes per day.	Souza-Alves et al. 2021b
Traveling (movement within the crown of a tree, between the crown of trees, shrubs, or on the ground that is not for the immediate purpose of finding food or engaging in social behavior), foraging (movements in a feeding site, e.g., crown of a feeding tree, cactus, ground, shrubs, for the purpose of obtaining food or in the case of plant exudates, gouging into tree bark), feeding (handling, manipulating, and ingesting of a food item, e.g., fruits, flowers, insects, vertebrates, exudates, seeds, and nectar), resting, social interactions.	Abreu et al. 2021
Locomotion (not defined), foraging for animals, feeding on plants (including manipulating them), feeding on animals, resting, social. Percentage for feeding includes both plants and animals.	Rylands 1989
Traveling (not defined), foraging for animal prey, feeding on exudates (including sporadic hole gouging), resting. Percentages for travel and foraging estimated from graphs.	Yepez et al. 2005
Moving (moving on substrate without any pausing to ingest food), foraging (actively moving on substrate, looking at the environment, occasionally eating insects, fruits, or other substance), feeding on fruit (feed exclusively on fruit from a single source), socializing, resting.	Matthews 2009
Travel (not defined), forage (all activities involved in finding and ingesting food), feed (consumption of plant materials), rest, miscellaneous.	Terborgh 1983
Travel, forage, rest, scan, social, other; not defined. Travel probably underestimated because of fast movement. Adult females only, estimated from a graph.	Rose 1994
Locomotion (any directional movement, including walking, jumping, galloping, climbing, descending, or running, whether within a tree, across trees, or on the ground), foraging (manual search for vegetable matter, among leaves or mature fruits, animal matter, manipulating items such as sticks and leaves, or a visual search for food), feeding (licking, biting, chewing, drinking, or ingesting dietary items), social interaction, resting or standing still, other.	de Oliveira et al. 2014
Locomote (move at least 1 m in 5 seconds), forage (behaviors directed toward a potential food source, including directed search, processing, and consuming), scan (visual inspection of surrounding area without fixed gaze; turning head side to side, scored while stationary or moving), social, rest. Combined foraging and feeding: 38%.	Fragaszy 1990

Scientific Name[1]	Common Name	Study Site	Travel (%)	Forage (%)	Feed (%)
Cebus olivaceus	Wedge-capped capuchin	Fundo Pecuario Masaguaral, Venezuela	22.2	28.2	18.4
Cephalopachus (Tarsius) bancanus	Western tarsier	Sepilok Forest Reserve, Sabah, Malaysia	26.5	60.1	2.1
Cercocebus atys	Sooty mangabey	Taï NP, Côte d'Ivoire	10.3	24.5	38.8
Cercopithecus ascanius	Redtailed monkey	Kakamega Forest, Kenya	40.3		
Cercopithecus ascanius	Redtailed monkey	Kibale Forest, Uganda	17.4	21.1	33.5
Cercopithecus campbelli	Campbell's monkey	Taï NP, Côte d'Ivoire	6.7	34.8	35.5
Cercopithecus diana	Diana monkey	Taï NP, Côte d'Ivoire	28.5	28.3	33.2
Cercopithecus diana	Diana monkey	Taï NP, Côte d'Ivoire	29.0—45.0	20.0—27.0	
Cercopithecus mitis	Blue monkey	Kakamega Forest, Kenya	13.1		
Cercopithecus mitis	Blue monkey	Kibale Forest, Uganda	16.5	8.9	33.3
Cercopithecus mitis	Blue monkey	Kibale Forest, Uganda	18.8	24.5	34.4

Activity categories and relevant notes	Source
Moving, foraging, resting, self-cleaning, grooming, playing, other; not defined. Graph differs in the categories: scanning also noted, and searching and processing, too. Percentage for "forage" combined from taking or attempting to take animal material (22.4%) and scanning (5.8%), percentage for "feed" combined from taking plant material (17%) and drinking (1.4%).	Robinson 1986
See Table 5.3.	Crompton and Andau 1986, 1987
See Table 5.3.	McGraw 1998
Locomote (any direction movement such as walking, running, or jumping not immediately associated with foraging), feed and forage (manipulation of potential food or inspection of microhabitats for invertebrate prey), rest, grooming, other. Activity budgets when species not in polyspecific association. Combined foraging and feeding: 37.5%.	Cords 1987
Climb, forage (manipulation of plant material in search of food, usually arthropod prey), feed (ingestion of food), scan (slow back and forth movement of the head; visual examination of substrate without manipulation), rest, self-clean, groom, cling, play, miscellaneous. Percentage for "Forage" in this table includes both "forage" and "scan."	Struhsaker and Leland 1979
See Table 5.3.	McGraw 1998
See Table 5.3.	McGraw 1998
Locomotion, foraging, resting, social; not defined. Percentages estimated from graphs.	Kane and McGraw 2018
Locomote (any direction movement such as walking, running, or jumping not immediately associated with foraging), feed and forage (manipulation of potential food or inspection of microhabitats for invertebrate prey), rest, grooming, other. Activity budgets when species not in polyspecific association. Combined foraging and feeding: 43.8%.	Cords 1987
Climb, forage (manipulation of plant material in search of food, usually arthropod prey), feed (ingestion of food), scan (slow back and forth movement of the head; visual examination of substrate without manipulation), rest, self-clean, groom, cling, play, miscellaneous. Percentage for "Forage" in this table includes both "forage" and "scan."	Struhsaker and Leland 1979
Climbing (not defined), foraging (manipulation of plant material in the apparent search for food), feeding (reaching for, picking, manipulating or placing food in the mouth), scanning (sitting or standing with head in constant motion while apparently searching nearby vegetation for food), resting, grooming, self-cleaning. Percentages averaged from six groups and are adult females only. Percentage for "Forage" in this table includes both "forage" and "scan."	Butynski 1990

Scientific Name[1]	Common Name	Study Site	Travel (%)	Forage (%)	Feed (%)
Cercopithecus mitis erythrarchus	Samango monkey	Cape Vidal, South Africa	22.6		35.8
Chiropotes sagulatus	Guianan bearded saki	Upper Essequibo Conservation Concession, Guyana	35.6		37.2
Chlorocebus djamdjamensis	Bale monkey	Odobullu Forest, Bale Mountains, Ethiopia	14.1		65.7
Chlorocebus djamdjamensis	Bale monkey	Odobullu Forest, Bale Mountains, Ethiopia	24.7		55.5
Chlorocebus pygerythrus (Cercopithecus aethiops)	Vervet monkey	Amboseli NP, Kenya	11.0—20.0		30.0—40.0
Chlorocebus pygerythrus (Cercopithecus aethiops)	Vervet monkey	Segera Ranch, Kenya	6.8	7.2	16.6
Chlorocebus pygerythrus	Vervet monkey	Mawana Game Reserve, S. Africa	9.0—13.0	38.0—50.0	
Chlorocebus sabaeus (Cercopithecus sabaeus)	Green monkey	Parc National du Niokolo-Koba, Senegal			44.8
Chlorocebus tantalus (Cercopithecus aethiops tantalus)	Tantalus monkey	Kala Maloue NP, Cameroon	125—150 min		190—200 min
Colobus angolensis palliatus	Angolan black-and-white colobus	Diani Forest, Kenya	5.7	1.5	24.7

Cercopithecus mitis erythrarchus–Colobus angolensis palliatus

Activity categories and relevant notes	Source
Moving (moving the entire body, including climbing, walking, jumping, and move scan, defined as moving slowly and attentive to a nearby potential food source with no manipulation), feeding (reaching for, picking, manipulating, or placing food in the mouth or manipulating the contents of a cheek pouch), resting (body stationary and not involved in activity, also including sit scanning, defined as attentive scanning of a potential food source with no manual disturbance of that food source), grooming or social (self-grooming and allo-grooming), play (exclusive to immatures).	Lawes and Piper 1992
Traveling (movement between tree crowns), moving (vertical or horizontal movement within a single tree crown that was not food related), feeding, resting, social behavior, other.	Shaffer 2013
Moving (changing spatial position, including walking, jumping, or running), feeding (manipulate, masticate, or ingest a food item), resting, playing, aggression, grooming, sexual.	Mekonnen et al. 2010
Moving (any locomotor behavior), feeding (foraging for or masticating food items), resting, socializing, vocalizing. Percentages are averages for two groups in non-fragmented forest only.	Mekonnen et al. 2017
Moving (any form of locomotion), feeding (manipulation or ingestion of food), scanning, allo-grooming, self-grooming, resting, other. Percentages are ranges estimated from graph of six groups.	Isbell and Young 1993a
See Table 5.3.	Isbell et al. 1998a, unpublished data
Moving (running, climbing, jumping, walking while not foraging), foraging (searching, reaching biting, chewing, licking, drinking), resting, social. Percentage ranges from four groups estimated from graph.	Canteloup et al. 2020
Feeding (foraging, picking, manipulating, chewing), resting, socializing.	Harrison 1985
Moving (walking, running, climbing, including manipulating grasses and visually scanning in search of mobile animals), feeding (manually or orally manipulating and ingesting food items including moving and resting for less than 30 seconds; manipulating a substrate in search of stationary or cryptic animals), day-resting, allo-grooming, night-resting. Values are minutes for one female estimated from graphs.	Nakagawa 2000
Move (not defined), forage (feeding while moving), feed (not defined), rest, social, other. Percentages are values for adults only in three groups.	Dunham 2015

Scientific Name[1]	Common Name	Study Site	Travel (%)	Forage (%)	Feed (%)
Colobus angolensis ruwenzorii	Rwenzori Angolan black-and-white colobus	Nabugabo, Uganda	25.0		28.0
Colobus angolensis ruwenzorii	Rwenzori Angolan black-and-white colobus	Nyungwe Forest, Rwanda	20.4		41.8
Colobus guereza	Eastern black-and-white colobus	Kibale Forest NP, Uganda	5.4		19.9
Colobus guereza	Eastern black-and-white colobus	Kibale Forest NP, Uganda	5.4		19.9
Colobus guereza	Eastern black-and-white colobus	Kakamega Forest, Kenya	1.8 and 2.8		28.3 and 22.9
Colobus guereza	Eastern black-and-white colobus	Kibale Forest NP, Uganda	6.4		16.5
Colobus polykomos	Western black-and-white colobus	Tiwai Island, Sierra Leone	11.6		
Colobus polykomos	Western black-and-white colobus	Taï NP, Côte d'Ivoire	15.1	10.8	34.9
Colobus satanas	Black colobus	Douala-Edea Reserve, Cameroon	3.6		22.5
Colobus vellerosus	Ursine colobus	Boabeng-Fiema Monkey Sanctuary, Ghana	14.6		23.7
Colobus vellerosus	Ursine colobus	Boabeng-Fiema Monkey Sanctuary, Ghana	6.8		22.0
Erythrocebus patas	Patas monkey	Segera Ranch, Kenya	8.6	21.5	15.0
Erythrocebus patas	Patas monkey	Kala Maloue NP, Cameroon	125—175 min		150—225 min

Colobus angolensis ruwenzorii–Erythrocebus patas

Activity categories and relevant notes	Source
Moving, feeding, resting, social, other; not defined.	Arseneau-Robar et al. 2021
Move (any locomotor behavior resulting in a change in spatial position), feed (pluck, masticate, or swallow food), rest, groom, aggression, other.	Fashing et al. 2007
Moving, feeding, play, inactivity, social grooming, self-cleaning; not defined.	Oates 1977a, b
Feed (ingestion of food), forage (manipulation of plant material in search of food, usually arthropod prey), scan (slow back and forth movement of the head; visual examination of substrate without manipulation), rest, climb (all forms of locomotion), self-clean, groom, cling, play, miscellaneous.	Struhsaker and Leland 1979
Move (locomotor behavior that resulted in change in spatial position), feed (pluck food items, pull food items toward mouth, masticate, swallow), rest, peer, groom, greet, social play, locomotor play. Percentages are averages for two groups.	Fashing 2001b
Moving, feeding, inactive, social grooming, self-cleaning, other; not defined.	Onderdonk and Chapman 2000
Travel, feed, rest, social, miscellaneous; not defined. Combined foraging and feeding: 30.8%	Dasilva 1992
See Table 5.3.	McGraw 1998
Locomotion (not defined), feeding (plucking, handling, ingesting food), sitting, lying, self-cleaning and grooming, clinging, playing, other.	McKey and Waterman 1982
Move, feed (manipulation and ingestion of food), rest, social. Percentages averaged from three groups.	Teichroeb et al. 2003
Moving, feeding, resting, social; not defined. Percentages averaged from three groups.	Wong and Sicotte 2007
See Table 5.3.	Isbell et al. 1998a, unpublished data
Moving (walking, running, climbing, including manipulating grasses and visually scanning in search of mobile animals), feeding (manually or orally manipulating and ingesting food items, including moving and resting for less than 30 seconds; manipulating a substrate in search of stationary or cryptic animals), day-resting, allo-grooming, night-resting. Values are minutes from one female estimated from graphs.	Nakagawa 2000

Scientific Name[1]	Common Name	Study Site	Travel (%)	Forage (%)	Feed (%)
Eulemur macaco	Black lemur	Ampasikely, Madagascar	32.0		15.0, 16.0, 23.0
Eulemur rubriventer	Red-bellied lemur	Ranomafana NP, Madagascar	22.0		20.0
Eulemur rubriventer	Red-bellied lemur	Ranomafana NP, Madagascar	52.8		34.5
Eulemur rufifrons (E. fulvus rufus)	Red-fronted brown lemur	Ranomafana NP, Madagascar	30.0		20
Eulemur rufifrons (E. fulvus rufus)	Red-fronted brown lemur	Ranomafana NP, Madagascar	52.4		34.2
Galago moholi (Galago senegalensis)	Southern lesser galago	Mosdene, Transvaal, South Africa	25.0	60.0	4.0
Gorilla beringei beringei	Mountain gorilla	Volcanoes NP, Rwanda			40.0
Gorilla beringei beringei	Mountain gorilla	Bwindi Impenetrable NP, Uganda	12.7		48.2
Gorilla beringei beringei	Mountain gorilla	Parc National des Volcans, Rwanda; Parc des Virungas, Zaire	6.5		55.4
Gorilla gorilla gorilla	Western lowland gorilla	Mondika Research Center, Central African Republic and Republic of Congo			43.8
Gorilla gorilla gorilla	Western lowland gorilla	Bai Hokou, Dzanga-Ndoki NP, Central African Republic	11.7		67.1
Gorilla gorilla gorilla	Western lowland gorilla	Maya Nord, Parc National d'Odzala, Republic of Congo	16.5		72.0
Hapalemur griseus	Gray bamboo lemur	Vatoharanana, Ranomafana NP, Madagascar	26.0		29.8
Hoolock hoolock	Western hoolock gibbon	Hollongapar Gibbon Wildlife Sanctuary, Assam, India			35.2

Eulemur macaco–Hoolock hoolock

Activity categories and relevant notes	Source
See Table 5.3.	Bayart and Simmen 2005
See Table 5.3.	Overdorff 1996
Travel (movement between trees), feed or forage (movement within a tree or a clump of contiguous, related trees, e.g., a grove of guava trees), rest, play. Percentages based on all locomotor bouts.	Dagosto 1995
See Table 5.3.	Overdorff 1996
Travel (movement between trees), feed or forage (movement within a tree or a clump of contiguous, related trees, e.g., a grove of guava trees), rest, play. Percentages based on all locomotor bouts.	Dagosto 1995
See Table 5.3.	Crompton 1984
Traveling (not defined), feeding (preparation and ingestion of food, including chewing but not time spent traveling while searching for food), resting, grooming, playing.	Grueter et al. 2016
Traveling (walking, running, climbing not directly related to foraging), feeding (foraging, food processing, and chewing) resting. Percentages are mean monthly values for adult females only.	Ostrofsky and Robbins 2020
Moving (locomotor activity other than shifts of position of less than 1 m during feeding, resting, and play), feeding (preparation and ingestion of food), resting, social, other.	Watts 1988
Travel, feed, rest; not defined. Percentages for adult females only.	Doran-Sheehy et al. 2009
Traveling (walking, running, climbing, not directly related to foraging), feeding (foraging, food processing, chewing), resting, social, other.	Masi et al. 2009
Locomotion, feeding, rest, social, visual surveillance, miscellaneous; not defined.	Magliocca and Gautier-Hion 2002
Travel or move (group movement, including changing position or moving less than 2 m), feed (search for, handle, and process food items), rest, social. Percentages averaged from dry and wet seasons.	Overdorff et al. 1997
Feeding (consumption of food plants species and other food items). Feeding calculated only.	Borah et al. 2018

Scientific Name[1]	Common Name	Study Site	Travel (%)	Forage (%)	Feed (%)
Hoolock leuconedys	Eastern hoolock gibbon	Nankang Park, Gaoligongshan National Nature Reserve, Yunnan, China	25.1		31.6
Hylobates agilis	Agile gibbon	Sungai Dal, Malaysia	30.0		36.0
Hylobates funereus	East Bornean gray gibbon	Danum Valley, Sabah, Malaysian Borneo	23.6—33.2		~21.0—32.0
Hylobates lar	Lar gibbon	Kuala Lompat, Peninsular Malaysia	32.0		42.0
Hylobates moloch	Javan gibbon	Gunung Halimun-Salak NP, Indonesia	15.0		36.0
Lagothrix lagotricha	Lowland woolly monkey	Yasuní NP, Ecuador	34.5	17.2	19.0
Lagothrix lagotricha	Lowland woolly monkey	Tinigua NP, Colombia	24.0 (3 years: 26.0)		36.0 (3 years: 35.0)
Lagothrix lagotricha	Woolly monkey	Estación Biológica Caparú, Department of Vaupés, Colombia	38.8	25.8	
Lagothrix lagotricha	Lowland woolly monkey	Tinigua NP, Colombia	24.0		36.0
Lemur catta	Ringtailed lemur	Anja Special Reserve; Tsaranoro Valley, Madagascar	10.0, 12.0		30.0, 49.0
Lemur catta	Ringtailed lemur	Berenty Reserve, Madagascar	17.0	8.5	14.1
Leontocebus nigrifrons (Saguinus fuscicollis)	Geoffroy's saddle-back tamarin	Rio Blanco Research Station, Peru	24.8	15.2	12.9
Leontocebus weddelli (Saguinus fuscicollis)	Weddell's saddle-back tamarin	Cocha Cashu, Manu NP, Peru	20.0	16.0	16.0

Hoolock leuconedys–Leontocebus weddelli

Activity categories and relevant notes	Source
Traveling (bipedal walk, suspension, leap, bridge, and climb), feeding (picking, chewing, or swallowing), resting, grooming, calling, social playing, other.	Fan et al. 2013
Travel, feed, rest, groom, sing, play, sleep; not defined.	Gittins and Raemakers 1980
Move (move and travel), feed (fig versus non-fig fruit), feed (leaves, flowers, or others), sing, play, groom, rest. Percentages from text and graph; minimum and maximum estimates. Percentages from one adult female.	Inoue et al. 2016
Travel, rest, groom, feed, sing, play, sleep; not defined.	Gittins and Raemakers 1980
Traveling, feeding, resting, socializing, intergroup aggression, other; not defined.	Kim et al. 2011
Move (change positions, either within or between tree crowns, exclusive of that movement taking place incidentally while searching a substrate), forage (search for animal prey while stationary or moving, to actively manipulate a substrate in search of prey, e.g., unrolling leaves, breaking apart branches, or inspecting vine tangles, or to actively attempt to procure prey items from the substrate), eat (handle, process, or consume either plant food items or animal prey), rest alone, rest social, social activity, other nonsocial.	Di Fiore and Rodman 2001
Moving, feeding on fruit, feeding on vegetative parts (leaves, stems, petioles, roots, and flowers), feeding on arthropods (larvae and adult arthropods), feeding on other items (termitaria soil and vertebrate prey), resting, other.	Stevenson et al. 1994
Move (any mobile activity, including walking, running, climbing, jumping, swinging, or brachiating within the same tree or between different trees), foraging (handing, ingestion, chewing, or obvious search for foods), rest, social, non-social.	Defler 1995
Moving, feeding, resting, social; not defined.	Stevenson et al. 2000
Locomote, forage (feed, drink), rest, social, territorial; not defined.	Gabriel 2013
Traveling, foraging, feeding, social activity, resting, miscellaneous; not defined.	Simmen et al. 2010
See Table 5.3.	Garber 1993
Travel (not defined), forage (all activities involved in finding and ingesting food), feed (consumption of plant materials), rest, miscellaneous.	Terborgh 1983

Scientific Name[1]	Common Name	Study Site	Travel (%)	Forage (%)	Feed (%)
Leontocebus weddelli (Saguinus fuscicollis)	Weddell's saddle-back tamarin	San Sebastian, Department of the Pando, Bolivia	22.0	10.0	8.0
Leontopithecus chrysomelas	Golden-headed lion tamarin	Una Biological Reserve, Bahia, Brazil	33.0		
Leontopithecus chrysomelas	Golden-headed lion tamarin	Lemos Maia Experimental Station, Cocoa Research Station, Bahia, Brazil	43.0	13.0	24.0
Leontopithecus rosalia	Golden lion tamarin	Reserva Biológica de Poça das Antas, Rio de Janeiro State, Brazil	33.5	19.7	18.4
Lepilemur leucopus	White-footed sportive lemur	Beza Mahafaly Special Reserve, Madagascar	12.5		33.0
Lepilemur sahamalaza	Sahamalaza sportive lemur	Ankarafa Forest, Sahamalaza-Iles Radama NP, Madagascar	12.2		9.7
Lepilemur sahamalaza	Sahamalaza sportive lemur	Ankarafa Forest, Sahamalaza-Iles Radama NP, Madagascar	3.6		18.7
Lophocebus albigena (Cercocebus albigena)	Gray-cheeked mangabey	Kibale Forest, Uganda	17.2	~10.0	32.5
Lophocebus albigena (Cercocebus albigena)	Gray-cheeked mangabey	Kibale Forest, Uganda	22.2	7.4	36.6
Lophocebus albigena	Gray-cheeked mangabey	Dja Reserve, Cameroon	27.0	15.0	40.0

Leontocebus weddelli–Lophocebus albigena

Activity categories and relevant notes	Source
Travel (quadrupedal movement including walking and running, branch-to-branch leaping, vertical clinging and leaping, and vertical ascent and descent), feed (eating or manipulating a food item), forage (searching for a food item within a substrate and visual search of the ground or plant for a food item), rest, other. Percentages from adult females in two groups.	Porter 2004
Traveling (not defined), eating animals, foraging for and eating fruit or flowers, foraging for animals, socializing, resting, remaining stationary, other. Combined foraging and feeding: 30.0%.	Raboy and Dietz 2004
Locomotion (not defined), foraging and feeding on animals, feeding on plants (including manipulating) resting, social. Percentage for feeding includes both plants and animals.	Rylands 1989
Travel (not defined), forage (manually searching for or eating animal prey items, included probing woody crevices, reaching down between leaf axis of bromeliads, sifting through dead leaves in palm crowns, ripping apart bark of trees and lianas, and grabbing, biting, and turning over rotting materials), feed (handling or chewing food of plant origin), rest, long call, intergroup interactions, scent mark, other. Percentages averaged from four groups except "feed" (two groups).	Dietz et al. 1997
Travel (move within or, more usually, between crowns), feed (pick, handle, ingest, or clearly chew food), rest, other. Percentages averaged medians from a cool, dry and a warm, wet season.	Nash 1998
Locomotion (walking, climbing, or jumping over a distance of >50 cm), movement of branches and leaves at the animal's location if out of view, feeding (handling food and eating, chewing visibly or audibly, rustling or dropping half-eaten food items), resting vigilant, resting, grooming, not visible, other. Percentages averaged from four seasons.	Mandl et al. 2018
Locomotion (climbs or jumps up or down a tree, jumps to another tree, or movements of vegetation indicate that focal animal advances), feeding (eating or processing food, biting and chewing noises heard at the focal's location), resting, vigilance, other, out of sight. Percentages averaged medians for seven females.	Seiler et al. 2014
See Table 5.3.	Waser 1977
Climb (all forms of locomotion), forage (manipulation of plant material in search of food, usually arthropod prey), feed (ingestion of food), scan (slow back and forth movement of the head; visual examination of substrate without manipulation), rest, self-clean, groom, cling, play, miscellaneous.	Struhsaker and Leland 1979
Traveling (walking, running, climbing, leaping), search (for insects; manipulation of a substrate in search of insects), feeding (manipulating a food item and bringing it to the mouth), resting, social behavior, other. Percentages estimated from graph.	Poulsen et al. 2001

Scientific Name[1]	Common Name	Study Site	Travel (%)	Forage (%)	Feed (%)
Lophocebus albigena (Cercocebus albigena)	Gray-cheeked mangabey	Kibale Forest, Uganda	28.0 and 2.07	41.5 and 40.7	
Lophocebus albigena johnstonii	Gray-cheeked mangabey	Kibale NP, Uganda	59.0	39.0	
Lophocebus albigena johnstonii	Gray-cheeked mangabey	Kibale NP, Uganda	29.0	43.0	
Lophocebus albigena	Gray-cheeked mangabey	Kibale NP, Uganda	29.0	4.7	28.8
Loris lydekkerianus (Loris tardigradus lydekkerianus)	Mysore slender loris	Ayyalur Interface Forestry Division, Tamil Nadu, India	22.3	27.0	0.9
Macaca assamensis	Assamese macaque	Nonggang Nature Reserve, Guanxi, China	34.0		20.0
Macaca assamensis	Assamese macaque	Phu Khieo Wildlife Sanctuary, Thailand	29.2		20.7
Macaca assamensis	Assamese macaque	Nagarujun forest, Shivapuri-Nagarjun NP, Nepal	20.0		55.0
Macaca assamensis	Assamese macaque	Askot, western Himalayas, Uttara-khand, India			38.0
Macaca assamensis	Assamese macaque	Guangxi Nonggang National Nature Reserve, Guangxi, China	28.6		32.7
Macaca fascicularis	Long-tailed macaque	Kuala Lompat, Krau Game Reserve, Malaysia	20.0		35.0
Macaca fascicularis	Long-tailed macaque	Ketambe Research Station, Sumatra, Indonesia	18.6	31.1	15.8

Lophocebus albigena–Macaca fascicularis

Activity categories and relevant notes	Source
Moving, foraging, resting; not defined. Two groups.	Olupot et al. 1994
Traveling, foraging, drinking, grooming; not defined. Males only. Values are for multiple study groups.	Janmaat and Chancellor 2010
Traveling, foraging, drinking, drinking and foraging (together 14%); not defined. Females only; values are for multiple study groups.	Janmaat and Chancellor 2010
Moving (horizontal or vertical displacement within or between trees), foraging (searching for invertebrates by tearing dead bark, breaking dead branches and twigs, or searching in and under epiphytes), feeding (reaching for, manipulating, and placing food items in the mouth), grooming, scanning (stationary, not feeding or obviously involved in a social interaction, facing somewhere other than the substrate immediately in front of it, and moving its head slowly), resting, other. Only males who were with groups of females; solitary males not included.	Olupot and Waser 2001
See Table 5.3.	Nekaris 2001
Moving, feeding, resting, social behavior; not defined. Percentages estimated from graph.	Zhou et al. 2014
Travel (based on positional behavior instead of activity), feed (ingesting, chewing, reaching for, or handling food items; feeding from cheek pouches not recorded as feeding), rest, social, other. Travel and feeding percentages not strictly comparable because of different ways of calculating them.	Heesen et al. 2013
Moving, feeding, resting, social; not defined. Feeding while moving or resting counted as feeding.	Koirala et al. 2017
Moving, feeding, resting, social, other; not defined.	Justa et al. 2019
Moving, feeding, resting, grooming, playing, other; not defined.	Li et al. 2020
Moving, feeding, resting, grooming, other; not defined.	Aldrich-Blake 1980
See Table 5.3.	van Schaik et al. 1983a

Scientific Name[1]	Common Name	Study Site	Travel (%)	Forage (%)	Feed (%)
Macaca fuscata	Japanese macaque	Yakushima Island, Japan	22.6		30.8
Macaca fuscata	Japanese macaque	Kinkazan Island, Japan	16.8		53.9
Macaca mulatta	Rhesus macaque	Nonggang Nature Reserve, Guanxi, China	28.0		38.0
Macaca mulatta	Rhesus macaque	Borme Village, Gazipur, Bangladesh	11.0		36.2
Macaca mulatta	Rhesus macaque	Gendaria-Shutrapur, Dhaka, Bangladesh	10.8		22.4
Macaca mulatta	Rhesus macaque	Askot, western Himalayas, Uttarakhand, India	24.0		
Macaca munzala	Arunachal macaque	Zemithang Valley, Arunachal Pradesh, India	19.0	29.0	
Macaca nigra	Crested macaque	Tangkoko-Duasaudara Nature Reserve, Sulawesi, Indonesia	22.5	12.7	23.6
Macaca silenus	Lion-tailed macaque	Varagaliyar Forest, Anamalai Wildlife Sanctuary, India	15.0	26.7	27.8
Macaca silenus	Lion-tailed macaque	Puthuthotam Cardamom Forest, Western Ghats, India	34.0	23.7	17.9
Macaca sylvanus	Barbary macaque	Oued El Abid, Central High Atlas Mountains, Morocco	17.0	19.0	26.0
Macaca sylvanus	Barbary macaque	Djurdjura NP; Akfadou, Algeria	22.3 and 20.0	3.9 and 6.2	23.8 and 25.4

Activity categories and relevant notes	Source
Moving, feeding, resting, social grooming, other; not defined.	Agetsuma and Nakagawa 1998
Moving, feeding, resting, social grooming, other; not defined.	Agetsuma and Nakagawa 1998
Moving, feeding, resting, social behavior; not defined. Percentages estimated from graph.	Zhou et al. 2014
Moving, feeding (includes foraging for and actual ingestion of food), resting, grooming, object manipulation, play, vigilance, dominance interactions.	Jaman and Huffman 2013
Moving, feeding (includes foraging for and actual ingestion of food), resting, grooming, object manipulation, play, vigilance, dominance interactions.	Jaman and Huffman 2013
Moving, feeding, resting, social, other; not defined.	Justa et al. 2019
Moving (not defined), foraging (not defined), sitting, resting, auto-grooming, alarm-calling, social.	Kumar et al. 2007
Moving (locomotion, including walking, running, climbing, and jumping), foraging (moving slowly with attention directed toward a potential food source or manipulating substrates in search of potential foods), feeding (reaching for, picking, manipulating, masticating, or placing food in mouth or manipulating the contents of a cheek). Percentages averaged from three groups.	O'Brien and Kinnaird 1997
Movement (lasting to count of 15 or involved in going to another level-zone), foraging (active search for and selection of foods before eating), feeding (actual eating), resting, sitting, standing, social, other.	Kurup and Kumar 1993
Ranging (directional travel, including movements from one canopy level to another or over long distances), foraging (active search for and selection of food items, including local travel within the canopy while searching for food items), feeding (manipulation or intake of food items), resting, other.	Menon and Poirier 1996
Moving (walking, running, climbing, and jumping), foraging (searching for, preparing, and handling food items), feeding (manipulating, placing food in mouth or chewing, but feeding from the cheek pouches excluded because it occurred mainly while monkeys rested or moved, and was therefore scored as such), resting, aggressive display. Percentages from non-provisioned group only and estimated from a graph.	El Alami et al. 2012
Moving (climbing, jumping, walking, running; observed almost exclusively between feeding or foraging bouts, so assumed greatest amount of moving time contributed to foraging effort), foraging (searching for food items before feeding, including turning over a stone and digging into the soil or under dead leaves, manipulation of plant material such as cleaning, hunting insects), feeding (picking and actual eating of food items), resting, social.	Ménard and Vallet 1997

Scientific Name[1]	Common Name	Study Site	Travel (%)	Forage (%)	Feed (%)
Macaca thibetana	Tibetan macaque	Tianhu Mountain County Nature Reserve, Mount Huangshan, Southern Anhui, China	27.0	28.6	
Microcebus murinus	Gray mouse lemur	Ankarafantsikia NP, Madagascar	11.8	17.5	30.2
Microcebus ravelobensis	Golden-brown mouse lemur	Ankarafantsikia NP, Madagascar	17.5	30.0	15.8
Mirza zaza	Giant northern mouse lemur	Ankarafa Forest, Sahamalaza, Madagascar	65.3		
Nasalis larvatus	Proboscis monkey	Menanggul River forests, Sabah, Malaysia	3.5		19.5
Nomascus concolor	Western black-crested gibbon	Dazhaizi, Mt. Wuliang, Yunnan, China	19.9		35.1
Nomascus concolor	Western black-crested gibbon	Bajiaohe, Yunnan, China	32.0		19.0
Nomascus concolor	Western black-crested gibbon	Dazhaizi Gibbon Research Center, Mt. Wuliang, Yunnan, China	23.8 and 28.8		24.5 and 26.9
Nomascus gabriellae	Yellow-cheeked crested gibbon	Cat Tien NP, Vietnam	14.1		45.0
Nomascus nasutus	Eastern black-crested gibbon	Bangliang Nature Reserve, Guangxi, China	18.8 and 30.0		23.9
Nycticebus bengalensis	Bengal slow loris	Trishna Wildlife Sanctuary, Tripura, India	23.0	6.2	22.3
Nycticebus bengalensis	Bengal slow loris	Phnom Samkos Wildlife Sanctuary, Cambodia	36.0		6.0
Nycticebus bengalensis	Bengal slow loris	Satchari NP, Bangladesh	42.2		17.2

Activity categories and relevant notes	Source
Moving (any behavior that causes a change in the individual's position, such as walking, climbing, jumping, and running more than 5 seconds), foraging (searching for food, handling food, and finally eating food), resting, grooming, other (included climbing and bridging).	Zhou et al. 2022
Moving (locomotion), foraging (searching for food), feeding, resting. When difficult to distinguish between foraging and moving, recorded as moving. Percentages averaged medians across seasons for females only.	Thorén et al. 2011
Moving (locomotion), foraging (searching for food), resting. When difficult to distinguish between foraging and moving, recorded as moving, feeding, resting. Percentages averaged medians across seasons for females only.	Thorén et al. 2011
See Table 5.3. Combined feeding and foraging: 14.6%.	Rode-Margono et al. 2016
Moving (locomotor behaviors resulting in a change in spatial position), feeding (handling, masticating, or swallowing food items), resting, other.	Matsuda et al. 2009
Traveling (movement including brachiating, climbing, jumping), feeding (picking, chewing, or swallowing food), calling, playing, other.	Fan et al. 2008
Moving (suspension, bipedal walking, leaping, or climbing), feeding (picking, handling, chewing, or swallowing food), resting, singing, other.	Ni et al. 2015
Moving, feeding, resting, singing, playing, other; not defined. Percentages from two groups.	Ning et al. 2019
Traveling (brachiating, leaping, bridging, bipedal walking, quadrupedal climbing, and walking with additional temporary use of the arms), feeding (looking for food and water, eating, and drinking), resting, social.	Bach et al. 2017
Traveling (bipedal walk, suspension, leap, bridge, climb), feeding (picking, chewing, or swallowing food), resting, grooming, calling, playing, other. Percentages from two groups, which had same percentage for feeding but not traveling.	Fan et al. 2012
Locomotion, foraging, feeding, grooming, resting, socializing; not defined.	Swapna et al. 2010
Moving (any mobile activity), feeding (consuming animal or plant matter), resting, sleeping, alert, other.	Rogers and Nekaris 2011
Traveling, feeding, hanging, resting, sleeping, grooming, other; not defined.	Al-Razi et al. 2020

Scientific Name[1]	Common Name	Study Site	Travel (%)	Forage (%)	Feed (%)
Nycticebus bengalensis	Bengal slow loris	Hollongapar Gibbon Wildlife Sanctuary, Assam, India	27.5		21.3
Nycticebus coucang	Sunda slow loris	Manjung District, Perak, West Malaysia			21.0
Nycticebus javanicus	Javan slow loris	Cipaganti, Cisurupan, Garut District, West Java	18.3	15.7	10.0
Nycticebus javanicus	Javan slow loris	Cipaganti, West Java	14.0		
Oreonax flavicauda	Yellow-tailed woolly monkey	Comunidad Capesina de Yambrasbamba, La Esperanza, Peru	29.0		5.0
Otolemur crassicaudatus (*Galago crassicaudatus*)	Thick-tailed greater galago	Wallacedale, Soutpansberg Mountains, Transvaal, South Africa	50.0	20.0	29.8
Pan paniscus	Bonobo	Lomako, Equateur, Zaire	16.1		40.4
Pan troglodytes schweinfurthii	Eastern chimpanzee	Mahale Mountains NP, Tanzania	15.4	21.3	
Pan troglodytes schweinfurthii	Eastern chimpanzee	Mahale Mountains NP, Tanzania	25.9—31.7		26.4—31.1
Pan troglodytes schweinfurthii	Eastern chimpanzee	Ngogo, Kibale NP, Uganda	14.0		47.0
Pan troglodytes schweinfurthii	Eastern chimpanzee	Kanyawara, Kibale NP, Uganda	11.0		44.0
Pan troglodytes schweinfurthii	Eastern chimpanzee	Gombe NP, Tanzania	13.8		55.7
Pan troglodytes verus	Western chimpanzee	Taï NP, Côte d'Ivoire	12.0		43.0

Nycticebus bengalensis–Pan troglodytes verus

Activity categories and relevant notes	Source
Locomotion, feeding, resting, alert, sleeping, auto-grooming, other, not defined.	Das and Nekaris 2020
Feeding (swallowing, chewing, or bringing animal prey or plant material to the mouth), resting, social interaction, other.	Wiens and Zitzmann 2003
Traveling, foraging, feeding, resting; not defined. Percentages averaged from three groups.	Reinhardt et al. 2016
Feeding or foraging, sleeping or resting, alert or freezing, social activity, grooming, other; not defined. Combined foraging and feeding: 31.0%.	Rode-Margono et al. 2014
See Table 5.3.	Crompton 1984
Traveling (change positions exclusive of incidental movement whilst foraging or engaged in another behavior [either within or between trees]), feeding (forage for, handle, process, or consume any food item [either plant part or animal prey]), resting, vocalizing, auto-grooming, allo-grooming, aggression, sexual, play, watching observer, out of sight.	Shanee and Shanee 2011
Traveling, feeding, other; not defined.	White 1992
Traveling, foraging, resting, social grooming, others; not defined.	Inoue and Shimada 2020
Moving (all locomotor activity other than shifts in position of less than 1 m during feeding and resting), feeding (preparation and ingestion of food), grooming, resting. Percentages averaged from anestrous and estrous females only.	Matsumoto-Oda and Oda 1998
Traveling (sustained movements greater than 1 min, generally occurring outside of feeding patches and involving movement between successive patches), feeding or foraging (ingestion or chewing of plant or animal matter uninterrupted by other behaviors for greater than or equal to 1 minute), resting, hunting, border patrolling.	Potts et al. 2011
Traveling (sustained movements greater than 1 minute, generally occurring outside of feeding patches and involving movement between successive patches), feeding and foraging (ingestion or chewing of plant or animal matter uninterrupted by other behaviors for greater than or equal to 1 minute), resting, hunting, border patrolling.	Potts et al. 2011
Travel, eat, allo-groom, rest, lie, not observed; not defined. Adult males only.	Wrangham 1977
Travel, feed, rest, groom, miscellaneous; not observed; not defined.	Doran 1997

Scientific Name[1]	Common Name	Study Site	Travel (%)	Forage (%)	Feed (%)
Pan troglodytes verus	Western chimpanzee	Taï NP, Côte d'Ivoire	22.0		45.0
Papio anubis	Olive baboon	Awash Valley, Ethiopia	25.0		30.2
Papio anubis	Olive baboon	Gilgil, Kenya	28.7		55.7
Papio cynocephalus	Yellow baboon	Amboseli NP, Kenya			
Papio hamadryas	Hamdryas baboon	Awash Valley, Ethiopia	25.3		29.5
Papio ursinus	Chacma baboon	Giant's Castle Reserve, Drakensberg Mountains, South Africa	25.0		50.0
Papio ursinus	Chacma baboon	Blyde Canyon Nature Reserve, Mpumalanga, South Africa	24.3	62.3	
Piliocolobus badius (Colobus badius)	Upper Guinea or Western red colobus	Taï NP, Côte d'Ivoire	18.9	15.8	29.1
Piliocolobus rufomitratus (Colobus badius rufomitratus)	Tana River red colobus	Tana River, Kenya	7.2		30.0
Piliocolobus tephrosceles (Procolobus rufomitratus)	Ashy red colobus	Kibale NP, Uganda	7.9		45.8
Piliocolobus tephrosceles (Colobus badius)	Ashy red colobus	Kibale Forest, Uganda	9.0	6.1	41.0
Piliocolobus tephrosceles (Procolobus rufomitratus tephrosceles)	Ashy red colobus	Kibale NP, Uganda	10.8		31.6

Pan troglodytes verus–Piliocolobus tephrosceles

Activity categories and relevant notes	Source
Travel, feed, meat-eating (9% in addition to the 45% for feeding), rest; not defined.	Boesch and Boesch-Achermann 2000
Moving (not defined), feeding (while sitting, while standing, and bipedally, including digging for food, feeding from the ground, bush, and tree), drinking, sitting, standing, grooming, other.	Nagel 1973
Travel, feed, rest, groom, social; not defined. Percentages from adult females only.	Eley et al. 1989
Moving (if both feeding and moving, coded as feeding), foraging (time spent feeding, moving, or doing both simultaneously), feeding, socializing, resting. Combined foraging and feeding: 60.0–71.0%.	Bronikowski and Altmann 1996
Moving (not defined), feeding (while sitting, while standing, and bipedally, including digging for food, feeding from the ground, bush and tree), drinking, sitting, standing, grooming, other.	Nagel 1973
Moving (walking more than 5 m without stopping to feed), foraging, resting, social. Percentages averaged from two groups and estimated from graph.	Whiten et al. 1987
Moving, foraging, resting, socializing; not defined.	Marais et al. 2006
See Table 5.3.	McGraw 1998
Moving, feeding, inactive, grooming (combined social and self-grooming), playing, clinging, other; not defined.	Marsh 1978
Traveling, feeding, resting; social (i.e., grooming, playing); not defined.	Gogarten et al. 2014
Climb (all forms of locomotion), forage (manipulation of plant material in search of food, usually arthropod prey), feed (ingestion of food), scan (slow back and forth movement of the head; visual examination of substrate without manipulation), rest, self-clean, groom, cling, play, miscellaneous. "Forage" in this table includes both "forage" and "scan."	Struhsaker and Leland 1979
Move, feed, inactive, social grooming, auto-grooming, other; not defined. Percentages are for adult females only and calculated from total sample size for both logging compartments.	Isbell 2012

Scientific Name[1]	Common Name	Study Site	Travel (%)	Forage (%)	Feed (%)
Piliocolobus tephrosceles (*Procolobus tephrosceles*)	Ashy red colobus	Kibale NP, Uganda	4.6—19.8		29.0—49.7
Pithecia irrorata	Gray's bald-faced saki	Los Amigos Conservation Concession, Madre de Dios, Peru	20.0		
Plecturocebus brunneus (*Callicebus moloch*)	Brown titi monkey	Manu NP, Peru	19.0	3.0	23.0
Plecturocebus cupreus (*Callicebus cupreus*)	Coppery titi monkey	Estación Biológica Quebrada Blanco, Peru	15.8	5.2	13.2
Plecturocebus discolor (*Callicebus discolor*)	Red-crowned titi monkey	Tiputini Biodiversity Station, Yasuní NP, Ecuador	27.0		14.5
Pongo abelii	Sumatran orangutan	Suaq Balimbing; Ketambe, Sumatra, Indonesia	15.1		56.3
Pongo abelii (*P. pygmaeus*)	Sumatran orangutan	Suaq Balimbing Research Station, Sumatra, Indonesia	17.0		55.0
Pongo pygmaeus	Bornean orangutan	Various locations, Borneo	14.1		48.3
Pongo pygmaeus morio	Northeast Bornean orangutan	Danum Valley Conservation Area, East Borneo, Sabah, Malaysia	16.6		50.6
Pongo pygmaeus morio	Northeast Bornean orangutan	Mentoko, Kutai Reserve, East Kalimantan, Indonesia	8.0—13.0		38.0—52.0
Pongo pygmaeus morio	Northeast Bornean orangutan	Kutai Reserve, East Kalimantan, Indonesia	11.1		45.9
Pongo pygmaeus wurmbii (*P. pygmaeus pygmaeus*)	Central Bornean orangutan	Tanjung Puting NP, Kalimantan Tengah, Borneo, Indonesia	18.7		60.1

Piliocolobus tephrosceles–Pongo pygmaeus wurmbii

Activity categories and relevant notes	Source
Travel, feeding, rest, social; not defined.	Chapman and Chapman 1999
Moving, resting, social, feeding and foraging; not defined. Combined foraging and feeding: 48.8%.	Palminteri and Peres 2012
Travel, forage (all activities involved in finding and ingesting food), feed (consumption of plant materials), rest, miscellaneous.	Terborgh 1983
Locomotion (moving a distance ≥1 m), feeding (eating pulp, seeds, leaves, flowers, prey, or other food items), foraging (looking for food, holding and manipulating food, grabbing prey). Percentages averaged from two groups.	Kulp and Heymann 2015
Moving (displacing self some distance, not in the context of a foraging bout), foraging (searching for, manipulating, or eating food items), feeding (takes food item into the mouth), resting, socializing, out of view. Percentages from adult females only and averaged from two groups; foraging percentages not reported.	Spence-Aizenberg et al. 2016
Traveling, feeding, resting, other; not defined. Percentages averaged from females only.	Morrogh-Bernard et al. 2009
Moving (not defined), feeding (gathering, manipulating, chewing, and swallowing food), resting, nesting, social, other.	Fox et al. 2004
Traveling, feeding, resting, other; not defined. Percentages averaged from females only.	Morrogh-Bernard et al. 2009
Traveling, feeding, resting, other; not defined. Percentages averaged from females only.	Kanamori et al. 2010
Traveling (all locomotor behavior with the exception of short movements during feeding), feeding (handling, processing, and swallowing of food items, as well as short movements within a feeding tree or area), resting, mating, vocalizing. Percentages are ranges for adult females only.	Mitani 1989
Moving, feeding, resting, nest-building, agonistic displaying; not defined.	Rodman 1979
Moving (not defined), foraging (inspect, reach for, prepare, extract, handle, bite, chew, or swallow food or moved within the tree or vine with the food source), resting, copulating, long-calling, agonistic, nesting.	Galdikas 1988

Scientific Name[1]	Common Name	Study Site	Travel (%)	Forage (%)	Feed (%)
Presbytis potenziani	Mentawai langur	Peleonan forest, North Siberut, Mentawai, West Sumatra, Indonesia	10.3	7.3	53.0
Presbytis potenziani	Mentawai langur	Betumonga, North Pagai Island, Mentawai, Indonesia	24.0		26.0
Presbytis rubicunda	Red langur	Natural Laboratory for the Study of Peat-Swamp Forest, Sagangau Forest, Central Kalimantan, Borneo, Indonesia	14.2		29.3
Presbytis thomasi	Thomas's langur	Bohorok, North Sumatra, Indonesia	9.0		28.2
Procolobus verus (Colobus) verus	Olive colobus	Taï NP, Côte d'Ivoire	19.1	12.7	26.5
Propithecus coronatus	Crowned sifaka	Antrema Station, Madagascar	<10.0	~1.0	35.0–45.0
Propithecus edwardsi (P. diadema edwardsi)	Milne-Edward's sifaka	Ranomafana NP, Madagascar	74.2		20.4
Pygathrix nemaeus	Red-shanked douc	Son Tra Nature Reserve, Vietnam	28.6		13.7
Rhinopithecus bieti	Yunnan snub-nosed monkey	Xiaochangdu, Honglaxueshan National Nature Reserve, Tibet	20.4		49.1
Rhinopithecus bieti	Yunnan snub-nosed monkey	Tacheng, Yunnan, China	15.0		35.0
Rhinopithecus bieti	Yunnan snub-nosed monkey	Xiangguqing, Baimaxueshan Nature Reserve, Yunnan, China	25.2		44.8
Rhinopithecus roxellana	Sichuan golden snub-nosed monkey	Laohegou Nature Reserve, Minshan Mountains, Sichuan, China	26.8	40.0	

Activity categories and relevant notes	Source
Traveling, foraging, feeding, resting, social behavior, other; not defined. Percentages calculated from the number of events per hour.	Hadi et al. 2012
Move, feed, rest, other; not defined.	Fuentes 1996
Traveling (locomotor behavior that resulted in a change in spatial position), feeding (manipulating, masticating or swallowing food items), resting, social behaviors, unknown.	Ehlers Smith et al. 2013
Moving, feeding, resting; not defined. Percentages averaged from two groups.	Gurmaya 1986
See Table 5.3.	McGraw 1998
Locomotion (traveling and locomotor activity other than foraging), foraging (movements that involve searching for food in tree crowns), feeding (food processing and chewing), resting, sleeping, sun-bathing, social, scent-marking, agonistic, miscellaneous. Range for feeding and percentage for locomotion is from the text for three groups. Percentage for foraging estimated from graph for all three groups.	Pichon and Simmen 2015
Travel (movement between trees), feed or forage (movement within a tree or a clump of contiguous, related trees, e.g., a grove of guava trees), rest, play. Percentages based on all locomotor bouts.	Dagosto 1995
Locomotion, feeding, inactivity, social; not defined.	Ulibarri and Gartland 2021
Moving (walking, running, or climbing), foraging (feeding + moving), feeding (ingestion, manipulation or inspection or chewing of food), non-foraging (resting + other), resting, other.	Xiang et al. 2010
Moving (walking, running, climbing), feeding (ingestion, manipulation, or inspection of food, and chewing), resting, social, other.	Ding and Zhao 2004
Moving (locomotion on the ground, bush, and tree for more than 5 seconds), feeding (feed from tree, bush, and ground; if moving, food must be ingested within 5 seconds), resting, grooming, other. Percentages from adult females only.	Li et al. 2014
Moving (walking on tree or swinging from tree to tree), foraging (searching for and eating food items), resting, other.	Fang et al. 2018

Scientific Name[1]	Common Name	Study Site	Travel (%)	Forage (%)	Feed (%)
Rhinopithecus roxellana	Sichuan golden snub-nosed monkey	Yuhuangmiao Village, Zhouzhi National Nature Reserve, Qinling Mountains, Shaanxi, China	22.9		35.8
Rhinopithecus roxellana	Sichuan golden snub-nosed monkey	Qianjiaping, Shennongjia National Nature Reserve, Hubei, China	26.2 and 30.5		30.3 and 33.9
Saguinus imperator	Emperor tamarin	Cocha Cashu, Manu NP, Peru	21.0	34.0	17.0
Saguinus labiatus	Red-bellied mustached tamarin	San Sebastian, Department of the Pando, Bolivia	31.0	12.0	10.0
Saguinus mystax	Moustached tamarin	Rio Blanco Research Station, Peru	31.4	14.4	13.1
Saimiri boliviensis (*Saimiri sciureus*)	Black-capped squirrel monkey	Cocha Cashu, Manu NP, Peru	27.0	50.0	11.0
Saimiri collinsi (*S. sciureus*)	Collins' squirrel monkey	Base 4 Wildlife Protection Zone; Germoplansma Island, Tucuruí Reservoir, Brazil	28.9	48.7 and 49.6	14.6 a nd 12.4
Saimiri collinsi	Collins' squirrel monkey	Ananim, Pará, Brazil	More	Less	18.0—22.0
Sapajus (*Cebus*) *apella*	Guianan brown capuchin	Nouragues Field Station, French Guiana	20.0—30.0	25.0—38.0	20.0—30.0
Sapajus (*Cebus*) *apella*	Guianan brown capuchin	Tinigua NP, Colombia	27.0		62.0
Sapajus cay	Azara's capuchin	Mato Grosso do Sul, Brazil	37.7	18.0	17.4
Sapajus libidinosus	Bearded capuchin	Fazenda Boa Vista, Brazil	32.0		

Rhinopithecus roxellana–Sapajus libidinosus

Activity categories and relevant notes	Source
Moving (traveling), feeding (actively manipulating potential food items, ingesting or masticating food), resting, other.	Guo et al. 2007
Move (any locational change, e.g., walking, jumping, and climbing), feeding (plucking or manipulating food items including insects under bark or rocks, using hands or mouth, or chewing food items; moving while feeding was feeding); resting, socializing. Percentages from adult females only in one large and one small group.	Liu et al. 2013
Travel (not defined), forage (all activities involved in finding and ingesting food), feed (consumption of plant materials), rest, miscellaneous.	Terborgh 1983
Travel (quadrupedal movement including walking and running, branch-to-branch leaping, vertical clinging and leaping, and vertical ascent and descent), forage (searching for a food item within a substrate, and visual search of the ground or plant for a food item), feed (eating or manipulating a food item), rest, other.	Porter 2004
See Table 5.3.	Garber 1988
Travel, forage (all activities involved in finding and ingesting food), feed (consumption of plant materials), rest, miscellaneous.	Terborgh 1983
Move, forage, feed, rest, social; not defined.	Pinheiro et al. 2013
Travel (direct movement not associated with foraging), eat (consumption of plant or animals), forage (visual or manual searches for plant or animal material), rest, social.	Ruivo et al. 2017
Move (rapid group progression and individual movement between tree crowns), forage (all activities involved in finding and ingesting animal matter), feed (consumption of vegetable matter, actual ingestion of items, including food preparation immediate before ingestion), rest, other. Percentages estimated from graph of monthly minimum and maximum estimates.	Zhang 1995
Moving, feeding, resting, social; not defined.	Stevenson et al. 2000
Traveling (moving from one place to another within the area, unaccompanied by any other form of activity), foraging (searching for food items, e.g., manipulating trunk, branches, or dry leaves, breaking branches in search of larvae, insects, fruits), feeding (ingesting food items, e.g., fruits, invertebrates, plant parts), resting, social interactions. Percentages for adult females only.	Júnior et al. 2019
Traveling, foraging (feeding included), resting, other; not defined. Combined foraging and feeding: 46.0%.	Izar et al. 2012

Scientific Name[1]	Common Name	Study Site	Travel (%)	Forage (%)	Feed (%)
Sapajus macrocephalus (Cebus apella)	Large-headed capuchin	Cocha Cashu, Manu NP, Peru	6.8	64.6	24.3
Sapajus macrocephalus (Cebus apella)	Large-headed capuchin	Cocha Cashu, Manu NP, Peru	21.0	50.0	16.0
Sapajus nigritus	Black-horned capuchin	Carlos Botelho State Park, Brazil	36.0		
Semnopithecus ajax	Kashmir gray langur	Dachigam NP, Kashmir, India	21.8		25.4
Semnopithecus schistaceus (S. entellus)	Nepal sacred langur	Langtang NP, Nepal	17.8		39.8
Semnopithecus entellus	Hanuman langur	Keshabpur; Manirampur, Jessore, Bangladesh	33.7		24.5
Semnopithecus hypoleucos	Malabar langur	Gerusoppa Forest Range, Karnataka, India	0.27		31.7
Semnopithecus johnii	Nilgiri langur	Nelliyampathy Reserve Forest, Kerala, India	0.77		26.7
Semnopithecus priam	Tufted gray langur	Pothigaiadi village, Western Ghats, India	13.8		23.3
Simias concolor	Pig-tailed langur	Peleonan forest, North Siberut Island, Mentawai, West Sumatra, Indonesia	9.3	3.7	46.2
Symphalangus (Hylobates) syndactylus	Siamang	Kuala Lompat, Peninsular Malaysia	22.0		50.0
Symphalangus (Hylobates) syndactylus	Siamang	Ketambe Research Station, Gunung Leuser Reserve, Sumatra, Indonesia	21.7		
Symphalangus syndactylus	Siamang	Way Canguk Research Area, Bukit Barisan Selatan NP, Sumatra, Indonesia	16.9		37.3
Tarsius spectrum	Spectral tarsier	Tangkoko Nature Reserve, Sulawesi, Indonesia		34.0—64.0	

Sapajus macrocephalus–Tarsius spectrum

Activity categories and relevant notes	Source
Travel, forage, feed, rest, other; not defined.	Janson 1985
Travel (not defined), forage (all activities involved in finding and ingesting food), feed (consumption of plant materials), rest, miscellaneous.	Terborgh 1983
Traveling, foraging (feeding included), resting, other; not defined. Combined foraging and feeding: 58.0%.	Izar et al. 2012
Moving (not defined), foraging (active intake of food and searching of food items), resting, social.	Mir et al. 2015
Travel, feed, rest, rest-huddle, rest-cling, groom, play, miscellaneous social behavior; not defined.	Sayers and Norconk 2008
Moving, feeding, resting, vigilance, grooming, playing, intergroup interactions; not defined. Rural group only (not urban group).	Khatun et al. 2015
Locomotion (moving from one place to another), feeding (foraging and eating), resting, self-directed, social.	Kavana et al. 2015
Locomotion (moving from one place to another), feeding (foraging and eating), resting, self-directed, social.	Kavana et al. 2015
Locomotion (running, chasing, jumping), feeding (collecting, chewing, or swallowing food), resting, social, conflict activity, other.	Vanaraj and Pragasan 2021
Traveling, foraging, feeding, resting, social behavior, other; not defined. Percentages calculated from the number of events per hour.	Hadi et al. 2012
Travel, feed, rest, groom, sing, play, sleep; not defined.	Gittins and Raemakers 1980
Traveling, foraging, resting, calling; not defined.	Karr 1982
Within-crown movement (movement within a feeding tree), travel (all other types of travel), feeding (reaching for, handling, chewing, or swallowing food), resting, social, other. Percentages from females only.	Lappan 2009
Traveling, foraging, resting, socializing; not defined. Percentage ranges span different moon phases and are from females only.	Gursky 2003

Scientific Name[1]	Common Name	Study Site	Travel (%)	Forage (%)	Feed (%)
Theropithecus gelada	Gelada	Indetu, Arsi, Ethiopia	20.3		41.7
Theropithecus gelada	Gelada	Gich, Simen Mountains NP, Ethiopia	14.7		62.3
Theropithecus gelada	Gelada	Sankaber, Simen Mountains NP, Ethiopia	20.4		45.2
Theropithecus gelada	Gelada	Bole, Simen Mountains NP, Ethiopia	17.4		35.7
Theropithecus gelada	Gelada	Guassa Community Conservation Area, Ethiopia	25.8		37.0
Trachypithecus crepusculus	Indochinese gray langur	Dazhaizi Research Station, Wuliangshan, Yunnan, China	31.9		20.8
Trachypithecus francoisi	François's langur	Nonggang Nature Reserve, Guangxi, China	17.3		23.1
Trachypithecus francoisi	François's langur	Mayanghe National Nature Reserve, Guizhou, China	7.0		29.0
Trachypithecus germaini	Indochinese silvered langur	Chua Hang Karst Mountain, Kien Giang, Vietnam	8.7		45.0
Trachypithecus leucocephalus	White-headed langur	Chongzuo White-headed Langur National Nature Reserve, Guangxi, China	18.4		24.2
Trachypithecus leucocephalus	White-headed langur	Fusui Rare and Precious Animal Nature Reserve, Guanxi, China	14.5		13.4
Trachypithecus leucocephalus	White-headed langur	Longlin, Fusui County, Guangxi, China	~7.0–13.0		9.0–20.0

Theropithecus gelada–Trachypithecus leucocephalus

Activity categories and relevant notes	Source
Moving (change in spatial position through locomotor behavior, including walking, jumping, or running), feeding (manipulate or masticate food item), resting, social activities.	Abu et al. 2018
Moving (quadrupedal locomotion), feeding (harvesting or passing to the mouth of any food matter), rest, social.	Iwamoto and Dunbar 1983
Moving (quadrupedal locomotion), feeding (harvesting or passing to the mouth of any food matter), rest, social.	Iwamoto and Dunbar 1983
Moving (quadrupedal locomotion), feeding (harvesting or passing to the mouth of any food matter), rest, social.	Iwamoto and Dunbar 1983
Moving, feeding, resting, playing, aggression, mating, grooming; not defined; data from one-male units only.	Mamo and Wube 2019
Traveling (movement within or between crowns of trees or on the ground in which the immediate purpose was neither to obtain a food item or interact with a conspecific), feeding (target individual grabs food item and manually processes, chews, or swallows it), resting, social behavior.	Pengfei et al. 2015
Moving (walking, running, jumping, or climbing posture that results in change of location; excludes movement during feeding), feeding (reaching for and manipulating a food item with hands or mouth, bringing it into the mouth and chewing; includes movement during feeding), grooming, playing, other.	Zhou et al. 2007
Moving (changing position or location), feeding (searching for food, lapping up water on leaves or pools, short-distance movements from branch to branch intently looking for food), grooming, nursing, resting, playing, guarding. Percentages estimated from graph.	Liu and Bhumpak-phan 2020
Traveling, feeding, resting, looking around, socializing; not defined.	Le et al. 2019
Moving (locomotor behavior resulting in a spatial position change, mainly walking, leaping, and climbing), feeding (manually or orally manipulating a food item), resting, grooming, playing, other. Percentages averaged from two groups.	Huang et al. 2017
Moving (moving from one site to another, without chasing, including changes of sitting places; also not eating after stopping), feeding (picking food off plants, putting food into the mouth, and chewing; also, moving, stopping, and then starting to feed), resting, playing, grooming, others.	Li and Rogers 2004
Moving (whole group locomotes forward or backward along a track, including walking, running, climbing, and jumping), feeding (search for or handle food items manually or orally), resting, sunbathing. Percentages are seasonal ranges.	Huang et al. 2003

Scientific Name[1]	Common Name	Study Site	Travel (%)	Forage (%)	Feed (%)
Trachypithecus poliocephalus	Cat Ba langur	Cat Ba Island, Vietnam	10.0	18.0	
Varecia rubra	Red ruffed lemur	Masoala NP, Madagascar	19.0		31.0
Varecia variegata	Black-and-white ruffed lemur	Ranomafana NP, Madagascar	9.0		29.5
Varecia variegata (*V. variegatus variegatus*)	Black-and-white ruffed lemur	Ranomafana NP, Madagascar	57.6		37.6

The search was conducted using the following Web of Science search terms: "activity budget* and (genus)," "time budget* and (genus)," "sex difference and primate* and activity budget*," "sex difference and monkey* and activity budget*," "sex difference and lemur* and activity budget*," "sex difference and ape* and activity budget*," "activity budget* and primate*," "activity budget* and monkey*," "activity budget* and lemur*," "activity budget* and ape*." Additional sources were found within searched publications.

1. Scientific names in parentheses are older names used in the cited studies.

2. NP is an abbreviation for National Park.

3. Definitions of feeding, foraging, moving, and locomotion in parentheses were taken verbatim from the cited studies.

Trachypithecus poliocephalus–Varecia variegata

Activity categories and relevant notes	Source
Locomotion (quadrupedal walking or running, climbing, leaping, dropping, and arm swinging), foraging (manipulating, searching for, and ingesting food).	Hendershott et al. 2016
Traveling (between feeding trees or other sites), feeding (including traveling within feeding trees), resting. Percentages from females only.	Vasey 2005
Traveling (not defined), feeding (actively searching for and consuming food items), socializing, resting. Percentages averaged from two communities.	Beeby and Baden 2021
Travel (movement between trees), feed or forage (movement within a tree or a clump of contiguous, related trees, e.g., a grove of guava trees), rest, play. Percentages based on all locomotor bouts.	Dagosto 1995

REFERENCES

Abreu, F., Garber, P. A., Souto, A., Presotto, A., and Schiel, N. 2021. Navigating in a challenging semiarid environment: The use of a route-based mental map by a small-bodied neotropical primate. *Animal Cognition* 24:629–643.

Abu, K., Mekonnen, A., Bekele, A., and Fashing, P. J. 2018. Diet and activity patterns of Arsi geladas in low-elevation disturbed habitat south of the Rift Valley at Indetu, Ethiopia. *Primates* 59:153–161.

Agetsuma, N., and Nakagawa, N. 1998. Effects of habitat differences on feeding behaviors of Japanese monkeys: Comparison between Yakushima and Kinkazan. *Primates* 39:275–289.

Agostini, I., Holzman, I., and Di Bitetti, M. S. 2010. Ranging patterns of two syntopic howler monkey species (*Alouatta guariba* and *A. caraya*) in northeastern Argentina. *International Journal of Primatology* 31:363–381.

Agostini, I., Holzman, I., and Di Bitetti, M. S. 2012. Influence of seasonality, group size, and presence of a congener on activity patterns of howler monkeys. *Journal of Mammalogy* 93:645–657.

Aldrich-Blake, F. P. G. 1980. Long-tailed macaques. In *Malayan Forest Primates: Ten Years' Study in Tropical Rain Forest* (Chivers, D., ed.). Plenum Press, New York, pp. 147–165.

Alexander, R. D. 1974. The evolution of social behavior. *Annual Review of Ecology and Systematics* 5:325–383.

Ali, R. 1981. *The Ecology and Social Behaviour of the Agastyamali Bonnet Macaque (Macaca radiata diluta)*. PhD dissertation, University of Bristol, Bristol, UK.

Allee, W. C. 1926. Measurement of environmental factors in the tropical rain-forest of Panama. *Ecology* 7:273–302.

Altmann, J. 1990. Males go where the females are. *Animal Behaviour* 39:193–195.

Altmann, J. 2000. Models of outcome and process: Predicting the number of males in primate groups. In *Primate Males: Causes and Consequences of Variation in Group*

Composition (Kappeler, P. M., ed.). Cambridge University Press, Cambridge, UK, pp. 236–247.

Altmann, J., and Alberts, S. C. 2003. Variability in reproductive success viewed from a life-history perspective in baboons. *American Journal of Human Biology* 15:401–409.

Altmann, S. A. 1974. Baboons, space, time and energy. *American Zoologist* 14:221–248.

Al-Razi, H., Hasan, S., Ahmed, T., and Muzaffar, S. B. 2020. Home range, activity budgets and habitat use in the Bengal slow loris (*Nycticebus bengalensis*) in Bangladesh. In *Evolution, Ecology and Conservation of Lorises and Pottos* (Nekaris, K. A. I., and Burrows, A. M., eds.). Cambridge University Press, New York, pp. 193–203.

Andelman, S. J. 1986. Ecological and social determinants of cercopithecine mating patterns. In *Ecological Aspects of Social Evolution: Birds and Mammals* (Rubenstein, D. I., and Wrangham, R. W., eds.). Princeton University Press, Princeton, NJ, pp. 201–216.

Anderson, C. M. 1986. Predation and primate evolution. *Primates* 27:15–39.

Angedakin, S., and Lwanga, J. 2011. Changes in group size and composition of the blue monkeys (*Cercopithecus mitis stuhlmanni*) between 1984 and 2009 at Ngogo, Kibale National Park, Uganda. *African Journal of Ecology* 49:362–366.

Arlet, M.E., and Isbell, L.A. 2009. Variation in behavioral and hormonal responses of adult male gray-cheeked mangabeys (*Lophocebus albigena*) to crowned eagles (*Stephanoaetus coronatus*). *Behavioral Ecology and Sociobiology* 63:491–499.

Arseneau-Robar, T. J. M., Changasi, A. H., Turner, E., and Teichroeb, J. 2021. Diet and activity budget in *Colobus angolensis ruwenzorii* at Nabugabo, Uganda: Are they energy maximizers? *Folia Primatologica* 92:35–48.

Asensio, N., Brockelman, W. Y., Malaivijitnond, S., and Reichard, U. H. 2011. Gibbon travel paths are goal oriented. *Animal Cognition* 14:395–405.

Ashbury, A. M., Willems, E. P., Atmoko, S. S. U., Saputra, F., van Schaik, C. P., and van Noordwijk, M. A. 2020. Home range establishment and the mechanisms of philopatry among female Bornean orangutans (*Pongo pygmaeus wurmbii*) at Tuanan. *Behavioral Ecology and Sociobiology* 74:42. DOI: 10.1007/s00265-020-2818-1

Aureli, F., Schaffner, C. M., Boesch, C., Bearder, S. K., Call, J., Chapman, C. A., Connor, R., Di Fiore, A., Dunbar, R. I. M., Henzi, S. P., Holecamp, K., Korstjens, A. H., Layton, R., Lee, P., Lehmann, J., Manson, J. H., Ramos-Fernandez, G., Strier, K. B., and van Schaik, C. P. 2008. Fission-fusion dynamics: New research frameworks. *Current Anthropology* 49:627–654 (with comments).

Aureli, F., Schaffner, C. M., Verpooten, J., Slater, K., and Ramos-Fernandez, G. 2006. Raiding parties of male spider monkeys: Insights into human warfare? *American Journal of Physical Anthropology* 131:486–497.

Austad, S. N., and Fischer, K. E. 1991. Mammalian aging, metabolism, and ecology: Evidence from the bats and marsupials. *Journal of Gerontology* 46:B47–B53.

Bach, T. H., Chen, J., Hoang, M. D., Beng, K. C., and Nguyen, V. T. 2017. Feeding behavior and activity budget of the southern yellow-cheeked crested gibbons (*Nomascus gabriellae*) in a lowland tropical forest. *American Journal of Primatology* 79:e22667. DOI: 10.1002/ajp.22667

Balasubramaniam, K. N., Dittmar, K., Berman, C. M., Butovskaya, M., Cooper, M. A., Majolo, B., Ogawa, H., Schino, G., Thierry, B., and de Waal, F. B. M. 2012a. Hierarchical steepness and phylogenetic models: Phylogenetic signals in *Macaca*. *Animal Behaviour* 83:1207–1218.

Balasubramaniam, K. N., Dittmar, K., Berman, C. M., Butovskaya, M., Cooper, M. A., Majolo, B., Ogawa, H., Schino, G., Thierry, B., and de Waal, F. B. M. 2012b. Hierarchical steepness, counter-aggression, and macaque social style scale. *American Journal of Primatology* 74:915–925.

Barnard, C. J. 1980. Flock feeding and time budgets in the house sparrow (*Passer domesticus* L.). *Animal Behaviour* 28:295–309.

Barocas, A., Ilany, A., Koren, L., Kam, M., and Geffen, E. 2011. Variance in centrality within rock hyrax social networks predicts adult longevity. *PLoS ONE* 6:e22375. DOI: 10.1371/journal.pone.0022375

Barry, R. E., and Barry, L. M. 1996. Species composition and age structure of remains of hyraxes (Hyracoidea: Procaviidae) at nests of black eagles. *Journal of Mammalogy* 77:702–707.

Bayart, F., and Simmen, B. 2005. Demography, range use, and behavior in black lemurs (*Eulemur macaco macaco*) at Ampasikely, Northwest Madagascar. *American Journal of Primatology* 67:299–312.

Bearder, S. K. 1999. Physical and social diversity among nocturnal primates: A new view based on long term research. *Primates* 40:267–282.

Bearder, S. K., Ambrose, L., Harcourt, C., Honess, P., Perkin, A., Pimley, E., Pullen, S., and Svoboda, N. 2003. Species-typical patterns of infant contact, sleeping site use and social cohesion among nocturnal primates in Africa. *Folia Primatologica* 74:337–354.

Bearder, S. K., and Doyle, G. A. 1974. Field and laboratory studies of social organization in bushbabies (*Galago senegalensis*). *Journal of Human Evolution* 3:37–50.

Bearder, S. K., and Martin, R. D. 1980. The social organization of a nocturnal primate revealed by radio tracking. In *A Handbook on Biotelemetry and Radio Tracking* (Amlaner, C. J., Jr., and Macdonald, D. W., eds.). Pergamon Press, New York, pp. 633–648.

Bearder, S. K., Nekaris, K. A. I., and Buzzell, C. A. 2002. Dangers in the night: Are some nocturnal primates afraid of the dark? In *Eat or be Eaten: Predator Sensitive Foraging Among Primates* (Miller, L. E., ed.). Cambridge University Press, New York, pp. 21–43.

Bearder, S. K., Nekaris, K. A. I., and Curtis, D. J. 2006. A re-evaluation of the role of vision in the activity and communication of nocturnal primates. *Folia Primatologica* 77:50–71.

Bearder, S. K., and Svoboda, N. 2013. *Otolemur crassicaudatus* Large-eared greater galago (thick-tailed greater galago/bushbaby). In *The Mammals of Africa: Vol II Primates* (Butynski, T. M., Kingdon, J., and Kalina, J., eds.). Bloomsbury Publishing, London, pp. 409–413.

Bearder, S. K., and Svoboda, N. 2016. Large-eared greater galago. In *All the World's Primates* (Rowe, N., and Myers, M., eds.). Pogonias Press, Charlestown, RI, pp. 131–132.

Beeby, N., and Baden, A. L. 2021. Seasonal variability in the diet and feeding ecology of black-and-white ruffed lemurs (*Varecia variegata*) in Ranomafana National Park, southeastern Madagascar. *American Journal of Physical Anthropology* 174:763–775.

Benadi, G., Fichtel, C., and Kappeler, P. M. 2008. Intergroup relations and home range use in Verreaux's sifaka (*Propithecus verreauxi*). *American Journal of Primatology* 70:956–965.

Bennie, J. J., Duffy, J. P., Inger, R., and Gaston, K. J. 2014. Biogeography of time partitioning in mammals. *Proceedings of the National Academy of Sciences USA* 111:13727–13732.

Bertram, B. C. R. 1980. Vigilance and group size in ostriches. *Animal Behaviour* 28:278–286.

Bianchi, R. C., and Mendes, S. L. 2007. Ocelot (*Leopardus pardalis*) predation on primates in Caratinga Biological Station, Southeast Brazil. *American Journal of Primatology* 69:1173–1178.

Bicca-Marques, J. C., and Calegaro-Marques, C. 1994. Activity budget and diet of *Alouatta caraya*: an age-sex analysis. *Folia Primatologica* 63:216–220.

Blanchong, J. A., and Smale, L. 2000. Temporal patterns of activity of the unstriped Nile rat, *Arvicanthis niloticus*. *Journal of Mammalogy* 81:595–599.

Blanco, M. B., Dausmann, K. H., Faherty, S. L., and Yoder, A. D. 2018. Tropical heterothermy is "cool": the expression of daily torpor and hibernation in primates. *Evolutionary Anthropology* 27:147–161.

Boesch, C. 1994. Chimpanzees—red colobus: A predator-prey system. *Animal Behaviour* 47:1135–1148.

Boesch, C., and Boesch H. 1989. Hunting behaviour of wild chimpanzees in the Taï National Park. *American Journal of Physical Anthropology* 78:547–573.

Boesch, C., and Boesch-Achermann, H. 2000. *The Chimpanzees of the Taï Forest: Behavioural Ecology and Evolution*. Oxford University Press, Oxford, UK.

Boesch, C., Crockford, C., Herbinger, I., Wittig, R., Moebius, Y., and Normand, E. 2008. Intergroup conflicts among chimpanzees in Taï National Park: Lethal violence and the female perspective. *American Journal of Primatology* 70:519–532.

Boinski, S., Kauffman, l., Ehmke, E., Schet, S., and Vreedzaam, A. 2005. Dispersal patterns among three species of squirrel monkeys (*Saimiri oerstedii*, *S. boliviensis* and *S. sciureus*): I. Divergent costs and benefits. *Behaviour* 142:525–632.

Boinksi, S., Sughrue, K., Selvaggi, L., Quatrone, R., Henry, M., and Cropp, S. 2002. An expanded test of the ecological model of primate social evolution: competitive regimes and female bonding in three species of squirrel monkeys (*Saimiri oerstedii*, *S. boliviensis*, and *S. sciureus*). *Behaviour* 139:227–261.

Bonadonna, G., Torti, V., Sorrentino, V., Randrianarison, R. M., Zaccagno, M., Gamba, M., Tan, C. L., and Giacoma, C. 2017. Territory exclusivity and intergroup encounters in the indris (Mammalia: Primates: Indridae: *Indri indri*) upon methodological tuning. *European Zoology Journal* 84:238–251.

Bonadonna, G., Zaccagno, M., Torti, V., Valente, D., De Gregorio, C., Randrianarison, R. M., Tan, C., Gamba, M., and Giacoma, C. 2020. Intra- and intergroup spatial dynamics of a pair-living singing primate, *Indri indri*: A multiannual study of three indri groups in Maromizaha Forest, Madagascar. *International Journal of Primatology* 41:224–245.

Bonebrake, T. C., Rezende, E. L., and Bozinovic, F. 2020. Climate change and thermo-regulatory consequences of activity time in mammals. *American Naturalist* 196:45–56.

Borah, M., Devi, A., and Kumar, A. 2018. Diet and feeding ecology of the western hoolock gibbon (*Hoolock hoolock*) in a tropical forest fragment of Northeast India. *Primates* 59:31–44.

Borries, C., Savini T., and Koenig, A. 2011. Social monogamy and the threat of infanticide in larger mammals. *Behavioral Ecology and Sociobiology* 65:685–693.

Boyer, D., Ramos-Fernández, G., Miramontes, O., Mateos, J. L., Cocho, G., Larralde, H., Ramos, H., and Rojas, F. 2006. Scale-free foraging by primates emerges from their interactions with a complex environment. *Proceedings of the Royal Society B: Biological Sciences* 273:1743–1750.

Brain, C., and Mitchell, D. 1999. Body temperature changes in free-ranging baboons (*Papio hamadryas ursinus*) in the Namib Desert, Namibia. *International Journal of Primatology* 20:585–598.

Braune, P., Schmidt, S., and Zimmermann, E. 2005. Spacing and group coordination in a nocturnal primate, the golden brown mouse lemur (*Microcebus ravelobensis*): the role of olfactory and acoustic signals. *Behavioral Ecology and Sociobiology* 58:587–596.

Brockelman, W. Y. 2009. Ecology and the social system of gibbons. In *The Gibbons* (Whittaker, D., and Lappan, S., eds.). Springer, New York, pp. 211–239.

Brockelman, W. Y., Reichard, U., Treesucon, U., and Raemaekers, J. J. 1998. Dispersal, pair formation and social structure in gibbons (*Hylobates lar*). *Behavioral Ecology and Sociobiology* 42:329–339.

Bronikowski, A. M., and Altmann, J. 1996. Foraging in a variable environment: weather patterns and the behavioral ecology of baboons. *Behavioral Ecology and Sociobiology* 39:11–25.

Bronikowski, A. M., Cords, M., Alberts, S. C., Altmann, J., Brockman, D. K., Fedigan, L. M., Pusey, A., Stoinski, T., Strier, K. B., and Morris, W. F. 2016. Female and male life tables for seven wild primate species. *Scientific Data* 3:160006. DOI: 10.1038/sdata.2016.6

Bshary, R. 2007. Interactions between red colobus monkeys and chimpanzees. In *Monkeys of the Tai Forest: An African Community* (McGraw, W. S., Zuberbühler, K., and Noë, R., eds.). Cambridge University Press, Cambridge, UK, pp. 155–170.

Bshary, R., and Noë, R. 1997. Behaviour of red colobus monkeys in the presence of chimpanzees. *Behavioral Ecology and Sociobiology* 41:321–333.

Buchanan-Smith, H. M., Hardie, S. M., Caceres, C., and Prescott, M. J. 2000. Distribution and forest utilization of *Saguinus* and other primates of the Pando Department, northern Bolivia. *International Journal of Primatology* 21:353–379.

Burnham, D., Bearder, S. K., Cheyne, S. M., Dunbar, R. I. M., and Macdonald, D. W. 2012. Predation by mammalian carnivores on nocturnal primates: Is the lack of evidence support for the effectiveness of nocturnality as an antipredator strategy? *Folia Primatologica* 83:236–251.

Bustamante-Manrique, S., Botero-Henao, N., Castaño, J. H., and Link, A. 2021. Activity budget, home range and diet of the Colombian night monkey (*Aotus lemurinus*) in peri-urban forest fragments. *Primates* 62:529–546.

Butler, H. 1980. The homologies of the lorisoid internal carotid artery system. *International Journal of Primatology* 1:333–343.

Butynski, T. M. 1990. Comparative ecology of blue monkeys (*Cercopithecus mitis*) in high- and low-density subpopulations. *Ecological Monographs* 60:1–26.

Caillaud, D., Eckardt, W., Vecellio, V., Ndagijimana, F., Mucyo, J.-P., Hirwa, J.-P., and Stoinski, T. 2020. Violent encounters between social units hinder the growth of a high-density mountain gorilla population. *Science Advances* 6:eabo724. DOI: 10.1002/ajp.22265

Caillaud, D., Ndagijimana, F., Giarrusso, A. J., Vecellio, V., and Stoinski, T. S. 2014. Mountain gorilla ranging patterns: Influence of group size and group dynamics. *American Journal of Primatology* 76:730–746.

Caine, N. G., and Stevens C. 1990. Evidence for a "monitoring call" in red-bellied tamarins. *American Journal of Primatology* 22:251–262.

Campos, F. A., Bergstrom, M. L., Childers, A., Hogan, J. D., Jack, K. M., Melin, A. D., Mosdossy, K. N., Myers, M. S., Parr, N. A., Sargeant, E., Schoof, V. A. M., and Fedigan, L. M. 2015. Drivers of home range characteristics across spatiotemporal scales in a Neotropical primate, *Cebus capucinus*. *Animal Behaviour* 91:93–109.

Campos, F. A., and Fedigan, L. M. 2009. Behavioral adaptations to heat stress and water scarcity in white-faced capuchins (*Cebus capucinus*) in Santa Rosa National Park, Costa Rica. *American Journal of Physical Anthropology* 138:101–111.

Campos, F. A., and Fedigan, L. M. 2014. Spatial ecology of perceived predation risk and vigilance behavior in white-faced capuchins. *Behavioral Ecology* 25:477–486.

Candiotti, A., Coye, C., Ouattara, K., Petit, E. J., Vallet, D., Zuberbühler, K., and Lemasson, A. 2015. Female bonds and kinship in forest guenons. *International Journal of Primatology* 36:332–352.

Cannon, C. H., and Leighton, M. 1994. Comparative locomotor ecology of gibbons and macaques: selection of canopy elements for crossing gaps. *American Journal of Physical Anthropology* 93:505–524.

Cant, J. G. H. 1987. Positional behavior of female Bornean orangutans (*Pongo pygmaeus*). *American Journal of Primatology* 12:71–90.

Canteloup, C., Borgeaud, C., Webs, M., and van de Waal, E. 2020. The effect of social and ecological factors on the time budget of wild vervet monkeys. *Ethology* 125:902–913.

Careau, V., Morand-Ferron, J., and Thomas, D. 2007. Basal metabolic rate of Canidae from hot deserts to cold arctic climates. *Journal of Mammalogy* 88:394–400.

Caro, T. M. 1989. Missing links in predator and antipredator behavior. *Trends in Ecology and Evolution* 4:333–334.

Caro, T. M. 1995. Pursuit-deterrence revisited. *Trends in Ecology and Evolution* 10:500–503.

Caro, T. M. 2005. *Anti-Predator Defenses in Birds and Mammals.* University of Chicago Press, Chicago, IL.

Cartmill, M. 1992. New views on primate origins. *Evolutionary Anthropology* 1:105:111.

Caselli, C. B., and Setz, E. Z. F. 2011. Feeding ecology and activity pattern of black-fronted titi monkeys (*Callicebus nigrifrons*) in a semideciduous tropical forest of southern Brazil. *Primates* 52:351–359.

Chapman, C. 1988. Patch use and patch depletion by the spider and howling monkeys of Santa Rosa National Park, Costa Rica. *Behaviour* 150:99–116.

Chapman, C. A. 1990. Association patterns of spider monkeys: The influences of ecology and sex on social organization. *Behavioral Ecology and Sociobiology* 26:409–414.

Chapman, C. A., and Chapman, L. J. 1999. Implications of small scale variation in ecological conditions for the diet and density of red colobus monkeys. *Primates* 40:215–231.

Chapman, C. A., Chapman, L. J., and McLaughlin, R. L. 1989. Multiple central place foraging by spider monkeys: Travel consequences of using many sleeping sites. *Oecologia* 79:506–511.

Chapman, C. A., Schoof, V. A. M., Bonnell, T. R., Gogarten, J. F., and Calmé, S. 2015. Competing pressures on populations: Long-term dynamics of food availability, food quality, disease, stress and animal abundance. *Philosophical Transactions of the Royal Society B* 370:20140112. DOI: 10.1098/rstb.2014.0112

Chapman, C. A., Wrangham, R. W., and Chapman, L. J. 1995. Ecological constraints on group size: An analysis of spider monkey and chimpanzee subgroups. *Behavioral Ecology and Sociobiology* 36:59–70.

Charles-Dominique, P. 1977. Urine-marking and territoriality in *Galago alleni* (Waterhouse, 1837 Lorisoidea, Primates)—a field study by radio-telemetry. *Zeitschrift für Tierpsychologie* 43:113–138.

Charnov, E. L. 1993. *Life History Invariants: Some Explorations of Symmetry in Evolutionary Ecology.* Oxford University Press, Oxford, UK.

Charnov, E. L., and Berrigan, D. 1993. Why do female primates have such long lifespans and so few babies? *or* life in the slow lane. *Evolutionary Anthropology* 1:191–194.

Chaves, O. M., Stoner, K. E., and Arroyo-Rodríguez, V. 2011. Seasonal differences in activity patterns of Geoffroyi's spider monkeys (*Ateles geoffroyi*) living in continuous and fragmented forests in southern Mexico. *International Journal of Primatology* 32:960–973.

Cheney, D. L. 1981. Intergroup encounters among free-ranging vervet monkeys. *Folia Primatologica* 35:124–146.

Cheney, D. L. 1992. Intragroup cohesion and intergroup hostility: The relation between grooming distributions and intergroup competition among female primates. *Behavioral Ecology* 3:334–345.

Cheney, D. L., and Seyfarth, R. M. 1981. Selective forces affecting the predator alarm calls of vervet monkeys. *Behaviour* 76:25–60.

Cheney, D. L., and Seyfarth, R. M. 1983. Nonrandom dispersal in free-ranging vervet monkeys: social and genetic consequences. *American Naturalist* 122:392–412.

Cheney, D. L., Seyfarth, R. M., Fischer, J., Beehner, J., Bergman, T., Johnson, S. E., Kitchen, D. M., Palombit, R. A., Rendall, D., and Silk, J. B. 2004. Factors affecting reproduction and mortality among baboons in the Okavango Delta, Botswana. *International Journal of Primatology* 25:401–428.

Cheney, D. L., and Wrangham, R. W. 1987. Predation. In *Primate Societies* (Smuts, B. B., Cheney, D. L., Seyfarth, R. M., Wrangham, R. W., and Struhsaker, T. T., eds.). University of Chicago Press, Chicago, IL, pp. 227–239.

Chen-Kraus, C., Raharinoro, N. A., Lawler, R. R., and Richard, A. F. 2023. Terrestrial tree hugging in a primarily arboreal lemur (*Propithecus verreaux*): a cool way to deal with heat? *International Journal of Primatology* 44:178–191.

Chism, J., and Rowell, T. E. 1988. The natural history of patas monkeys. In *A Primate Radiation: Evolutionary Biology of the African Guenons* (Gautier-Hion, A., Bouliere F., Gautier, J.-P., and Kingdon, J., eds.). Cambridge University Press, New York, pp. 412–438.

Chivers, D. J. 1977. The feeding behaviour of siamang (*Symphalangus syndactylus*). In *Primate Ecology: Studies of Feeding and Ranging Behavior in Lemurs, Monkeys and Apes* (Clutton-Brock, T. H., ed.). Academic Press, New York, pp. 355–382.

Clark, A. B. 1985. Sociality in a nocturnal "solitary" prosimian: *Galago crassicaudatus*. *International Journal of Primatology* 6:581–600.

Clark, R. W. 2005. Pursuit-deterrent communication between prey animals and timber rattlesnakes (*Crotalus horridus*): The response of snakes to harassment displays. *Behavioral Ecology and Sociobiology* 59:258–261.

Clarke, A., Rothery, P., and Isaac, N. J. B. 2010. Scaling of basal metabolic rate with body mass and temperature in mammals. *Journal of Animal Ecology* 79:610–619.

Clarke, E., Reichard, U. H., and Zuberbühler, K. 2012. The anti-predator behaviour of wild white-handed gibbons (*Hylobates lar*). *Behavioral Ecology and Sociobiology* 66:85–96.

Clink, D. J., Dillis, C., Feilen, K. L., Beaudrot, L., and Marshall, A. J. 2017. Dietary diversity, feeding selectivity, and responses to fruit scarcity of two sympatric Bornean primates (*Hylobates albibarbis* and *Presbytis rubicunda rubida*). *PLoS ONE* 12:e0173369. DOI: 10.1371/journal.pone.0173369

Clutton-Brock, T. H. 1989. Mammalian mating systems. *Proceedings of the Royal Society B: Biological Sciences* 236:339–372.

Clutton-Brock, T. H., and Harvey, P. H. 1977a. Primate ecology and social organization. *Journal of Zoology, London* 183:1–39.

Clutton-Brock, T. H., and Harvey, P. H. 1977b. Sexual dimorphism, socionomic sex ratio and body weight in primates. *Nature* 269:797–800.

Clutton-Brock, T. H., and Lukas, D. 2012. The evolution of social philopatry and dispersal in female mammals. *Molecular Ecology* 21:472–492.

Cody, M. L. 1971. Finch flocks in the Mohave Desert. *Theoretical Population Biology* 2:142–158.

Coleman, B. T., and Hill, R. A. 2013. Living in a landscape of fear: The impact of predation, resource availability and habitat structure on primate range use. *Animal Behaviour* 88:165–173.

Colquhoun, I. C. 2011. A review and interspecific comparison of nocturnal and cathemeral strepsirhine primate olfactory behavioural ecology. *International Journal of Zoology* 2011:362976. DOI: 10.1155/2011/362976

Cook, M., and Mineka, S. 1989. Observational conditioning of fear to fear-relevant versus fear-irrelevant stimuli in rhesus monkeys. *Journal of Abnormal Psychology* 98:448–459.

Cords, M. 1987. Mixed-species association of *Cercopithecus* monkeys in the Kakamega Forest, Kenya. *University of California Publications in Zoology* 117:1–109.

Cords, M. 2002. Friendships among adult female blue monkeys (*Cercopithecus mitis*). *Behaviour* 139:291–314.

Cords, M. 2007. Variable participation in the defense of communal feeding territories by blue monkeys in the Kakamega Forest, Kenya. *Behaviour* 144:1537–1550.

Coss, R. G., McCowan, B., and Ramakrishnan, U. 2007. Threat-related acoustical differences in alarm calls by wild bonnet macaques (*Macaca radiata*) elicited by python and leopard models. *Ethology* 113:352–367.

Cowlishaw, G. 1994. Vulnerability to predation in baboon populations. *Behaviour* 131;293–304.

Cowlishaw, G. 1997. Alarm calling and implications for risk perception in a desert baboon population. *Ethology* 103:384–394.

Crockett, C. M. 1984. Emigration by female red howler monkeys and the case for female competition. In *Female Primates: Studies by Women Primatologists* (Small, M. F., ed.). Alan R. Liss, New York, pp. 159–173.

Crockett, C. M., and Eisenberg, J. F. 1987. Howlers: Variations in group size and demography. In *Primate Societies* (Smuts, B. B., Cheney, D. L., Seyfarth, R. M., Wrangham, R. W., and Struhsaker, T. T., eds.). University of Chicago Press, Chicago, IL, pp. 54–68.

Crockett, C. M., and Janson, C. H. 2000. Infanticide in red howlers: Female group size, group composition, and a possible link to folivory. In *Infanticide by Males and its Implications* (van Schaik, C. P., and Janson, C. H., eds). Cambridge University Press, Cambridge, UK, pp. 75–98.

Crockett, C. M., and Pope, T. R. 1988. Inferring patterns of aggression from red howler monkey injuries. *American Journal of Primatology* 15:289–308.

Crockett, C. M., and Pope, T. R. 1993. Consequences of sex differences in dispersal for juvenile red howler monkeys. In *Juvenile Primates: Life History, Development, and Behavior* (Pereira, M. E., and Fairbanks, L. A., eds.). Oxford University Press, New York, pp. 104–118.

Crockford, C., Wittig, R. M., Mundry, R., and Zuberbühler, K. 2012. Wild chimpanzees inform ignorant group members of danger. *Current Biology* 22:142–146.

Crompton, R. W. 1984. Foraging, habitat structure, and locomotion in two species of *Galago*. In *Adaptations for Foraging in Nonhuman Primates: Contributions to an Organismal Biology of Prosimians, Monkeys and Apes* (Rodman, P. S., and Cant, J. G. H., eds.). Columbia University Press, New York, pp. 73–111.

Crompton, R. H., and Andau, P. M. 1986. Locomotion and habitat utilization in free-ranging *Tarsius bancanus*: A preliminary report. *Primates* 27:337–355.

Crompton, R. H., and Andau, P. M. 1987. Ranging, activity rhythms, and sociality in free-ranging *Tarsius bancanus*: A preliminary report. *International Journal of Primatology* 8:43–71.

Crook, J. H. 1963. A comparative analysis of nest structure in the weaver birds (Ploceinae). *Ibis* 105:238–262.

Crook, J. H. 1965. The adaptive significance of avian social organizations. *Symposia of the Zoological Society of London* 14:181–218.

Crook, J. H., and Gartlan, J. C. 1966. Evolution of primate societies. *Nature* 210: 1200–1203.

Dagosto, M. 1995. Seasonal variation in positional behavior of Malagasy lemurs. *International Journal of Primatology* 16:807–833.

Dammhahn, M., and Kappeler, P. M. 2005. Social system of *Microcebus berthae*, the world's smallest primate. *International Journal of Primatology* 26:407–435.

Dammhahn, M., and Kappeler, P. M. 2008. Comparative feeding ecology of sympatric *Microcebus berthae* and *M. murinus*. *International Journal of Primatology* 29:1567–1589.

Dammhahn, M., and Kappeler, P. M. 2009. Females go where the food is: does the socio-ecological model explain variation in social organization of solitary foragers? *Behavioral Ecology and Sociobiology* 63:939–952.

Darwin, C. 1871. *The Descent of Man, and Selection in Relation to Sex*. John Murray, London.

Das, N., and Nekaris, K. A. I. 2020. Positional behaviour and substrate preference of slow lorises, with a case study of *Nycticebus bengalensis* in Northeast India. In *Evolution, Ecology and Conservation of Lorises and Pottos* (Nekaris, K. A. I., and Burrows, A. M., eds.). Cambridge University Press, New York, pp. 219–227.

Dasilva, G. L. 1992. The western black-and-white colobus as a low-energy strategist: Activity budgets, energy expenditure and energy intake. *Journal of Animal Ecology* 61:79–91.

Davidge, C. 1978. Ecology of baboons (*Papio ursinus*) at Cape Point. *Zoologica Africana* 13:329–350.

Defler, T. R. 1995. The time budget of a group of wild woolly monkeys (*Lagothrix lagotricha*). *International Journal of Primatology* 16:107–120.

de Guinea, M., Estrada, A., Janmaat, K. R. M., Nekaris, K. A.-I., and Van Belle, S. 2021. Disentangling the importance of social and ecological information in goal-directed movements in a wild primate. *Animal Behaviour* 173:41–51.

de Guinea, M., Estrada, A., Nekaris, K. A.-I., and Van Belle, S. 2019. Arboreal route navigation in a Neotropical mammal: Energetic implications associated with tree monitoring and landscape attributes. *Movement Ecology* 7:39 DOI: 10.1186/s40462-019-0187-z

de Jager, M., Bartumeus, F., Kölzsch, A., Swissing, F. J., Hengeveld, G. M., Noet, B. A., Herman, P. M. J., and van de Koppel, J. 2014. How superdiffusion gets arrested: Ecological encounters explain shift from Lévy to Brownian movement. *Proceedings of the Royal Society B: Biological Sciences* 281:20132605. DOI: 10.1098.rspb.2013.2605

de Oliveira, S. G., Lynch Alfaro, J. W., and Veiga, L. M. 2014. Activity budget, diet, and habitat use in the critically endangered ka'apor capuchin monkey (*Cebus kaapori*) in Pará State, Brazil: A preliminary comparison to other capuchin monkeys. *American Journal of Primatology* 76:919–931.

de Ruiter, J. 1986. The influence of group size on predator scanning and foraging behaviour of wedge-capped capuchin monkeys (*Cebus olivaceus*). *Behaviour* 98:240–258.

DeVore, I. 1963. Comparative ecology and behaviour of monkeys and apes. In *Classification and Human Evolution* (Washburn, S. L., ed.). Wenner Gren, New York, pp. 301–319.

DeVore, I., and Hall, K. R. L. 1965. Baboon ecology. In *Primate Behavior* (DeVore, I., ed.). Holt, Rinehart, and Winston, New York, pp. 20–52.

DeVore, I., and Lee, R. 1963. Recent and current field studies in primates. *Folia Primatologica* 1:66–72.

Dias, L. G., and Strier, K. B. 2003. Effects of group size on ranging patterns in *Brachyteles arachnoides hypoxanthus*. *International Journal of Primatology* 24:209–221.

Dias, P. A. D., Coyohua-Fuentes, A., Canales-Espinosa, D., and Rangel-Negrín, A. 2015. Group structure and dynamics in black howlers (*Alouatta pigra*): A 7-year perspective. *International Journal of Primatology* 36:311–331.

Dias, P. A. D., Rangel-Negrin, A., and Canales-Espinosa, D. 2011. Effects of lactation on the time-budgets and foraging patterns of female black howlers (*Alouatta pigra*). *American Journal of Physical Anthropology* 145:137–146.

Dietz, J. M., Peres, C. A., and Pinder, L. 1997. Foraging ecology and use of space in wild golden lion tamarins (*Leontopithecus rosalia*). *American Journal of Primatology* 41:289–305.

Di Fiore, A. 2003. Ranging behavior and foraging ecology of lowland woolly monkeys (*Lagothrix lagotricha poeppigii*) in Yasuní National Park, Ecuador. *American Journal of Primatology* 59:47–66.

Di Fiore, A., and Fleischer, R. C. 2005. Social behavior, reproductive strategies, and population genetic structure of *Lagothrix poeppigii*. *International Journal of Primatology* 26:1137–1173.

Di Fiore, A., Link, A., Schmitt, C. A., and Spehar, S. N. 2009. Dispersal patterns in sympatric woolly and spider monkeys: Integrating molecular and observational data. *Behaviour* 146:437–470.

Di Fiore, A., and Rendall, D. 1994. Evolution of social organization: A reappraisal for primates by using phylogenetic methods. *Proceedings of the National Academy of Sciences USA* 91:9941–9945.

Di Fiore, A., and Rodman, P. S. 2001. Time allocation patterns in lowland woolly monkeys (*Lagothrix lagotricha poeppigii*) in a Neotropical terra firma forest. *International Journal of Primatology* 22:449–480.

Di Fiore, A., and Suarez, S. A. 2007. Route-based travel and shared routes in sympatric spider and woolly monkeys: Cognitive and evolutionary implications. *Animal Cognition* 10:317–329.

Ding, E., and Zhao, Q.-K. 2004. *Rhinopithecus bieti* at Tacheng, Yunnan: Diet and daytime activities. *International Journal of Primatology* 25:583–598.

Dolotovskaya, S., Roos, C., and Heymann, E. W. 2020. Genetic monogamy and mate choice in a pair-living primate. *Scientific Reports* 10:20328. DOI: 10.1038/s41598-020-77132-9

Doolan, S. P., and Macdonald, D. W. 1996. Diet and foraging behaviour of group-living meerkats, *Suricata suricatta*, in the southern Kalahari. *Journal of Zoology, London* 239:697–716.

Doran, D. 1997. Influence of seasonality on activity patterns, feeding behavior, ranging, and grouping patterns in Taï chimpanzees. *International Journal of Primatology* 18:183–206.

Doran-Sheehy, D., Mongo, P., Lodwick, J., and Conklin-Brittain, N. L. 2009. Male and female western gorilla diet: Preferred foods, use of fallback resources, and implications for ape versus Old World monkey foraging strategies. *American Journal of Physical Anthropology* 140:727–738.

Drea, C. M., Goodwin, T. E., and delBarco-Trillo, J. 2019. P-Mail: the information highway of nocturnal, but not diurnal or cathemeral, strepsirrhines. *Folia Primatologica* 90:422–438.

Drösher, I., and Kappeler, P. M. 2013. Defining the low end of primate social complexity: the social organization of the nocturnal white-footed sportive lemur (*Lepilemur leucopus*). *International Journal of Primatology* 34:1225–1243.

Dunbar, R. I. M. 1988. *Primate Social Systems.* Cornell University Press, Ithaca, NY.

Dunbar, R. I. M., and Dunbar, E. P. 1974. Ecological relations and niche separation between sympatric terrestrial primates in Ethiopia. *Folia Primatologica* 21:36–60.

Dunham, N. T. 2015. Ontogeny of positional behavior and support use among *Colobus angolensis palliates* of the Diani Forest Kenya. *Primates* 56:183–192.

Eberle, M., and Kappeler, P. M. 2008. Mutualism, reciprocity, or kin selection? Cooperative rescue of a conspecific from a boa in a nocturnal solitary forager, the gray mouse lemur. *American Journal of Primatology* 70:410–414

Ehlers Smith, D. A., Ehlers Smith, Y. C., and Cheyne, S. M. 2013. Home-range use and activity patterns of the red langur (*Presbytis rubicunda*) in Sabangau tropical peat-swamp forest, Central Kalimantan, Indonesian Borneo. *International Journal of Primatology* 34:957–972.

Eisenberg, J. F., Muckenhirn, N. A., and Rudran, R. 1972. The relation between ecology and social structure in primates. *Science* 176:863–874.

Ekernas, L. S., and Cords, M. 2007. Social and environmental factors influencing natal dispersal in blue monkeys, *Cercopithecus mitis stuhlmanni. Animal Behaviour* 73:1009–1020.

El Alami, A., van Lavieren, E., Rachida. A., and Chait, A. 2012. Differences in activity budgets and diet between semiprovisioned and wild-feeding groups of the endangered Barbary macaque (*Macaca sylvanus*) in the Central High Atlas Mountains, Morocco. *American Journal of Primatology* 74:210–216.

Eley, R. M., Strum, S. C., Muchemi, G., and Reid, G. D. F. 1989. Nutrition, body condition, activity patterns, and parasitism of free-ranging troops of olive baboons (*Papio anubis*) in Kenya. *American Journal of Primatology* 18:209–219.

Ellefson, J. O. 1968. Territorial behaviour in the common white-handed gibbon, *Hylobates lar* Linn. In *Primates: Studies in Adaptation and Variability* (Jay, P., ed.). Holt, Rinehart, and Winston, New York, pp. 180–199.

Emery Thompson, M., Kahlenberg, S. M., Gilby, I. C., and Wrangham, R. W. 2007. Core area quality is associated with variance in reproductive success among female chimpanzees at Kibale National Park. *Animal Behaviour* 73:501–512.

Emlen, S. T., and Oring, L. W. 1977. Ecology, sexual selection, and the evolution of mating systems. *Science* 197:215–223.

Emmons, L. H. 2000. *Tupai: A Field Study of Bornean Treeshrews.* University of California Press, Berkeley.

Erhart, E. M., and Overdorff, D. J. 2008a. Spatial memory during foraging in prosimian primates: *Propithecus edwardsi* and *Eulemur fulvus rufus*. *Folia Primatologica* 79:185–196.

Erhart, E. M., and Overdorff, D. J. 2008b. Population demography and social structure changes in *Eulemur fulvus rufus* from 1988 to 2003. *American Journal of Physical Anthropology* 136:183–193.

Estrada, A., Garber, P. A, Rylands, A. B., Roos, C., Fernandez-Duque, E., Di Fiore, A., Nekaris, K. A. I., Nijman, V., Heymann, E. W., Lambert, J. E., Rovero, F., Barelli, C., Setchell, J. M., Gillespie, T. R., Mittermeier, R. A., Arregoitia, L. V., de Guinea, M., Gouveia, S., Dobrovolski, R., Shanee, S., Shanee, N., Boyle, S. A., Fuentes, A., MacKinnon, K. C., Amato, K. R., Meyer, A. L. S., Wich, S., Sussman, R. W., Pan, R., Kone, I., and Li, B. 2017. Impending extinction crisis of the world's primates: Why primates matter. *Science Advances* 3:e1600946. DOI: 10.1126/sciadv.1600946

Estrada, A., Juan-Solano, S., Martínez, T. O., and Coates-Estrada, R. 1999. Feeding and general activity patterns of a howler monkey (*Alouatta palliata*) troop living in a forest fragment at Los Tuxtlas, Mexico. *American Journal of Primatology* 48:167–183.

Fan, P.-F., Ai, H.-S., Fei, H.-L., Zhang, D., and Yan, S.-D. 2013. Seasonal variation of diet and time budget of Eastern hoolock gibbons (*Hoolock leuconedys*) living in a northern montane forest. *Primates* 54:137–146.

Fan, P.-F., Fe, H.-L., and Ma, C.-Y. 2012. Behavioral responses of Cao Vit gibbon (*Nomascus nasutus*) to variations in food abundance and temperature in Bangliang, Jingxi, China. *American Journal of Primatology* 74:632–641.

Fan, P.-F., and Jiang, X.-L. 2010. Maintenance of multifemale social organization in a group of *Nomascus concolor*) at Wuliang Mountain, Yunnan, China. *International Journal of Primatology* 31:1–13.

Fan, P.-F., Ni, Q.-Y., Sun, G.-S., Huang, B., and Jiang, X.-L. 2008. Seasonal variations in the activity budget of *Nomascus concolor jingdongensis* at Mt. Wuliang, Central Yunnan, China: Effects of diet and temperature. *International Journal of Primatology* 29:1047–1057.

Fang, G., Li, M., Liu, X.-J., Guo, W.-J., Jiang, Y.-T., Huang, Z.-P., Tang, S.-Y., Li, D.-Y., Yu, J., Jin, T., Lui, X.-G., Wang, J.-M., Li, S., Qi, X.-G., and Li, B.-G. 2018. Preliminary report on Sichuan golden snub-nosed monkeys (*Rhinopithecus roxellana roxellana*) at Laohegou Nature Reserve, Sichuan, China. *Scientific Reports* 8:16183. DOI: 10.1038/s41598-018-34311-z

Fashing, P. J. 2001a. Male and female strategies during intergroup encounters in guerezas (*Colobus guereza*): Evidence for resource defense mediated through males and a comparison with other primates. *Behavioral Ecology and Sociobiology* 50:219–230.

Fashing, P. J. 2001b. Activity and ranging patterns of guerezas in the Kakamega Forest: Intergroup variation and implications for intragroup feeding competition. *International Journal of Primatology* 22:549–577.

Fashing, P. J. 2011. African colobine monkeys: Their behavior, ecology, and conservation. In *Primates in Perspective, 2nd edition* (Campbell, C. J., Fuentes, A., MacKinnon, K. C., Bearder, S. K., and Stumpf, R. M., eds.). Oxford University Press, New York, pp. 203–229.

Fashing, P. J., Mulindahabi, F., Gakima, J.-B., Masozera, M., Mununura, I., Plumptre, A. J., and Nguyen, A. 2007. Activity and ranging patterns of *Colobus angolensis ruwenzorii* in Nyungwe Forest, Rwanda: Possible costs of large group size. *International Journal of Primatology* 28:529–550.

Fedigan, L. M., and Baxter, M. J. 1984. Sex differences and social organization in free-ranging spider monkeys (*Ateles geoffroyi*). *Primates* 25:279–294.

Fedigan, L. M., and Jack, K. 2001. Neotropical primates in a regenerating Costa Rican dry forest: A comparison of howler and capuchin population patterns. *International Journal of Primatology* 22:689–713.

Feinsinger, P. 1976. Organization of a tropical guild of nectarivorous birds. *Ecological Monographs* 46:257–291.

Felton, A. M., Felton, A., Lindenmayer, D. B., and Foley, W. J. 2009. Nutritional goals of wild primates. *Functional Ecology* 23:70–78.

Fernandez-Duque, E. 2009. Natal dispersal in monogamous owl monkeys (*Aotus azarai*) of the Argentinean Chaco. *Behaviour* 146:583–606.

Fernandez-Duque, E., and Erkert, H. G. 2006. Cathemerality and lunar periodicity of activity rhythms in owl monkeys of the Argentinian Chaco. *Folia Primatologica* 77:123–138.

Ferrari, S. F., and Hilário, R. R. 2014. Seasonal variation in the length of the daily activity period in buffy-headed marmosets (*Callithrix flaviceps*): An important consideration for the analysis of foraging strategies in observational field studies of primates. *American Journal of Primatology* 76:385–392.

Fichtel, C. 2004. Reciprocal recognition of sifaka (*Propithecus verreauxi verreauxi*) and redfronted lemur (*Eulemur fulvus rufus*) alarm calls. *Animal Cognition* 7:45–52.

Fichtel, C. 2007. Avoiding predators at night: Antipredator strategies in red-tailed sportive lemurs (*Leoplemur ruficaudatus*). *American Journal of Primatology* 69:611–624.

Fichtel, C., and Hilgartner, R. 2013. Noises in the dark: Vocal communication in *Lepilemur ruficaudatus* and other nocturnal pair-living primates. In *Leaping Ahead: Advances in Prosimian Biology, Developments in Primatology: Progress and Prospects* (Masters, J., Gamba, M., and Génin, F., eds.). Springer, New York, pp. 297–304.

Fleagle, J. G., and Mittermeier, R. A. 1980. Locomotor behavior, body size, and comparative ecology of seven Surinam monkeys. *American Journal of Physical Anthropology* 52:301–314.

Fogden, M. 1974. A preliminary field-study of the western tarsier, *Tarsius bancanus* Horsfield. In *Prosimian Biology* (Martin, R. D., Doyle, G. A., and Walker, A. C., eds.). Pittsburgh University Press, Pittsburgh, PA, pp. 151–165.

Ford, S. M. 1980. Callitrichids as phyletic dwarfs, and the place of the Callitrichidae in Platyrrhini. *Primates* 21:31–43.

Fox, E. A., van Schaik, C. P., Sitompul, A., and Wright, D. N. 2004. Intra- and interpopulational differences in orangutan (*Pongo pygmaeus*) activity and diet: Implications for the invention of tool use. *American Journal of Physical Anthropology* 125:162–174.

Fragaszy, D. M. 1980. Comparative studies of squirrel monkeys (*Saimiri*) and titi monkeys (*Callicebus*) in travel tasks. *Zeitschrift für Tierpsychologie* 54:1–36.

Fragaszy, D. M. 1990. Sex and age differences in the organization of behavior in wedge-capped capuchins, *Cebus olivaceus. Behavioral Ecology* 1:81–94.

Fragaszy, D. M., Visalberghi, E., and Fedigan, L. M. 2004. *The Complete Capuchin: the Biology of the Genus Cebus.* Cambridge University Press, Cambridge, UK.

Fuentes, A. 1996. Feeding and ranging in the Mentawai Island langur (*Presbytis potenziani*). *International Journal of Primatology* 17:525–548.

Fuller, A., Mitchell, D., Maloney, S. K., Hetem, R. S., Fonsêca, V. F. C., Meyer, L. C. R., van de Ven, T. M. F. N., and Snelling, E. P. 2021. How dryland mammals will respond to climate change: The effects of body size, heat load and a lack of food and water. *Journal of Experimental Biology* 224: jeb238113. DOI: 10.1242/jeb.238113

Gabriel, D. N. 2013. Habitat use and activity patterns as an indication of fragment quality in a strepsirrhine primate. *International Journal of Primatology* 34:388–406.

Galdikas, B. M. F. 1988. Orangutan diet, range, and activity at Tanjung Puting, Central Borneo. *International Journal of Primatology* 9:1–35.

Garber, P. A. 1988. Foraging decisions during nectar feeding by tamarin monkeys (*Saguinus mystax* and *Saguinus fuscicollis*, Callitrichidae, Primates) in Amazonian Peru. *Biotropica* 20:100–106.

Garber, P. A. 1993. Seasonal patterns of diet and ranging in two species of tamarin monkeys: Stability versus variability. *International Journal of Primatology* 14:145–166.

Garber, P. A, Moya, L., and Malaga, C. 1984. A preliminary field study of the moustached tamarin monkey (*Saguinus mystax*) in northeastern Peru: Questions concerned with the evolution of a communal breeding system. *Folia Primatologica* 42:17–32.

Garcia, J. E., and Braza, F. 1987. Activity rhythms and use of space of a group of *Aotus azarae* in Bolivia during the rainy season. *Primates* 28:337–342.

Gaulin, S. J. C., and Gaulin, C. K. 1982. Behavioral ecology of *Alouatta seniculus* in Andean cloud forest. *International Journal of Primatology* 3:1–32.

Gaulin, S. J. C., and Sailer, L. D. 1985. Are females the ecological sex? *American Anthropologist* 87:111–119.

Génin, F. G. S., Masters, J. C., and Ganzhorn, J. U. 2010. Gummivory in cheirogaleids: Primitive retention or adaptation to hypervariable environments? In *The Evolution of Exudativory in Primates* (Burrows, A. M., and Nash, L. T., eds.). Springer, New York, pp. 123–140.

Genoud, M. 2002. Comparative studies of basal rate of metabolism in primates. *Evolutionary Anthropology* 11(Supplement 1):108–111.

Genoud, M., Martin, R. D., and Glaser, D. 1997. Rate of metabolism in the smallest simian primate, the pygmy marmoset (*Cebuella pygmaea*). *American Journal of Primatology* 41:229–245.

Gerber, B. D, Arrigo-Nelson, S., Karpanty, S. M., Kotschwar, M., and Wright, P. C. 2012. Spatial ecology of the endangered Milne-Edwards' sifaka (*Propithecus edwardsi*): Do logging and season affect home range and daily ranging patterns? *International Journal of Primatology* 33:305–321.

Gerloff, U., Hartung, B., Fruth, B., Hohmann, G., and Tautz, D. 1999. Intracommunity relationships, dispersal pattern and paternity success in a wild living community of bonobos (*Pan paniscus*) determined from DNA analysis of faecal samples. *Proceedings of the Royal Society B: Biological Sciences* 266:1189–1195.

Gibson, L., and Koenig, A. 2012. Neighboring groups and habitat edges modulate range use in Phayre's leaf monkeys (*Trachypithecus phayrei crepusculus*). *Behavioral Ecology and Sociobiology* 66:633–643.

Gilbert, L. E. 1975. Ecological consequences of a coevolved mutualism between butterflies and plants. In *Coevolution of Animals and Plants* (Gilbert, L. E., and Raven, P. H., eds.). University of Texas Press, Austin, pp. 210–240.

Gilchrist, J. S. 2006. Female eviction, abortion, and infanticide in banded mongooses (*Mungos mungo*): Implications for social control of reproduction and synchronized parturition. *Behavioral Ecology* 17:664–669.

Gill, F. B. 1988. Trapline foraging by hermit hummingbirds: Competition for an undefended, renewable resource. *Ecology* 69:1933–1942.

Gittins, S. P., and Raemakers, J. J. 1980. Siamang, lar, and agile gibbons. In *Malayan Forest Primates: Ten Years' Study in Tropical Rain Forest* (Chivers, D., ed.). Plenum Press, New York, pp. 63–105.

Gleiss, A. C., Jorgensen, S. J., Leibsch, N., Sala, J. E., Norman, B., Hays, G. C., Quintana, F., Grundy, E., Campagna, C., Trites, A. W., Block, B. A., and Wilson, R. P. 2011. Convergent evolution in locomotory patterns of flying and swimming animals. *Nature Communications* 2:352. DOI: 10.1038/ncomms1350

Gogarten, J. F., Bonnell, T. R., Brown, L. M., Campenni, M., Wasserman, M. D., and Chapman, C. A. 2014. Increasing group size alters behavior of a folivorous primate. *International Journal of Primatology* 35:590–608.

Gogarten, J. F., Jacob, A. L., Ghai, R. R., Rothman, J. M., Twinomugisha, D., Wasserman, M. D., and Chapman, C. A. 2015. Group size dynamics over 15+ years in an African forest primate community. *Biotropica* 47:101–112.

Goodman, S. M., O'Connor, S., and Langrand, O. 1993. A review of predation on lemurs: Implications for the evolution of social behavior in small, nocturnal primates. In *Lemur Social Systems and Their Ecological Basis* (Kappeler, P. M., and Ganzhorn, J. U., eds.). Plenum Press, New York, pp. 51–66.

Gordon, A. D., Johnson, S. E., and Louis, E. E., Jr. 2013. Females are the ecological sex: Sex-specific body mass ecogeography in wild sifaka populations (*Propithecus* spp.). *American Journal of Physical Anthropology* 151:77–87.

Goss-Custard, J., Dunbar, R., and Aldrich-Blake, P. 1972. Survival, mating and rearing strategies in the evolution of primate social structure. *Folia Primatologica* 17:1–19.

Greenwood, P. J. 1980. Mating systems, philopatry, and dispersal in birds and mammals. *Animal Behaviour* 28:1140–1162.

Grange, S., Duncan, P., Gaillard, J.-M., Sinclair, A. R. E., Gogan, P. J. P., Packer, C., Hofer, H., and East, M. 2004. What limits the Serengeti zebra population? *Oecologia* 140:523–532.

Grueter, C. C., Ndamiyabo, F., Plumptre, A. J., Abavandimwe, D., Mundry, R., Fawcett, K. A., and Robbins, M. M. 2013. Long-term temporal and spatial dynamics of food availability for endangered mountain gorillas in Volcanoes National Park, Rwanda. *American Journal of Primatology* 75:267–280.

Grueter, C. C., Robbins, A. M., Abavandimwe, D., Vecellio, V., Ndagijimana, F., Ortmann, S., Stoinski, T. S., and Robbins, M. M. 2016. Causes, mechanisms, and consequences of contest competition among female mountain gorillas in Rwanda. *Behavioral Ecology* 27:766–776.

Guan, Z.-H., Ma, C.-Y., Fei, H.-L., Huang, B., Ning, W.-H., Ni, Q.-Y., Jiang, X.-L., and Fan, P.-F. 2018. Ecology and social system of northern gibbons living in cold seasonal forests. *Zoological Research* 39:255–265.

Guiden, P. W., and Orrrock, J. L. 2020. Seasonal shifts in activity timing reduce heat loss of small mammals during winter. *Animal Behaviour* 164:181–192.

Guo, S., Li, B., and Watanabe, K. 2007. Diet and activity budget of *Rhinopithecus roxellana* in the Qinling Mountains, China. *Primates* 48:268–276.

Gurmaya, K. J. 1986. Ecology and behavior of *Presbytis thomasi* in northern Sumatra. *Primates* 27:151–172.

Gursky, S. 2002. Predation on a wild spectral tarsier (*Tarsius spectrum*) by a snake. *Folia Primatologica* 73:60–62.

Gursky, S. 2003. Lunar philia in a nocturnal primate. *International Journal of Primatology* 24:351–367.

Gursky, S. 2005. Associations between adult spectral tarsiers. *American Journal of Physical Anthropology* 128:74–83.

Gursky, S. 2006. Function of snake mobbing in spectral tarsiers. *American Journal of Physical Anthropology* 129:601–608.

Gursky, S. 2010. Dispersal patterns in *Tarsius spectrum*. *International Journal of Primatology* 31:117–131.

Gursky, S., and Nekaris, K. A. I. 2019. Nocturnal primate communication: ecology, evolution and conservation. *Folia Primatologica* 90:273–278.

Hadi, S., Ziegler, T., Waltert, M., Syamsuri, F., Mühlenberg, M., and Hodges, J. K. 2012. Habitat use and trophic niche overlap of two sympatric colobines, *Presbytis potenziani* and *Simias concolor*, on Siberut Island, Indonesia. *International Journal of Primatology* 33:218–232.

Hall, K. R. L. 1965. Behaviour and ecology of the wild patas monkey, *Erythrocebus patas*, in Uganda. *Journal of Zoology, London* 148:15–87.

Hamilton, W. D. 1971. Geometry for the selfish herd. *Journal of Theoretical Biology* 31:295–311.

Hankerson, S. J., and Dietz, J. M. 2014. Predation rate and future reproductive potential explain home range size in golden lion tamarins. *Animal Behaviour* 96:87–95.

Harcourt, A. H. 1978. Strategies of emigration and transfer by primates: with particular reference to gorillas. *Zeitschrift für Tierpsychologie* 48:401–420.

Harcourt, A. H., and Stewart, K. J. 2007. *Gorilla Society: Conflict, Compromise, and Cooperation between the Sexes*. University of Chicago Press, Chicago, IL.

Harcourt, C. 1991. Diet and behaviour of a nocturnal lemur, *Avahi laniger*, in the wild. *Journal of Zoology, London* 223:667–674.

Harrison, J. S. 1985. Time budget of the green monkey, *Cercopithecus sabaeus*: some optimal strategies. *International Journal of Primatology* 6:351–376.

Hart, D. L., and Sussman, R. W. 2005. *Man the Hunted: Primates, Predators, and Human Evolution*. Westview Press, Boulder, CO.

Hartwell, K. S., Notman, H., Bonenfant, C., and Pavelka, M. S. M. 2014. Assessing the occurrence of sexual segregation in spider monkeys (*Ateles geoffroyi yucatanensis*), its mechanisms and function. *International Journal of Primatology* 35:425–444.

Hasegawa, T. 1990. Sex differences in ranging patterns. In *The Chimpanzees of the Mahale Mountains: Sexual and Life History Strategies* (Nishida, T., ed.). University of Tokyo Press, Tokyo, pp. 99–114.

Hasson, O. 1991. Pursuit-deterrent signals: Communication between prey and predator. *Trends in Ecology and Evolution* 6:325–329.

Healy, K., Guillerme, T., Finlay, S., Kane, A., Kelly, S. B. A., McClean, D., Kelly, D. J., Donohue, I., Jackson, A. L., and Cooper, N. 2014. Ecology and mode-of-life explain lifespan variation in birds and mammals. *Proceedings of the Royal Society B: Biological Sciences* 281:20140298. DOI: 10.1098/rspb.2014.0298

Heesen, M., Rogahn, S., Ostner, J., and Schülke, O. 2013. Food abundance affects energy intake and reproduction in frugivorous female Assamese macaques. *Behavioral Ecology and Sociobiology* 67:1053–1066.

Heinrich, B. 1976. The foraging specializations of individual bumblebees. *Ecological Monographs* 46:105–128.

Hendershott, R., Behie, A., and Rawson, B. 2016. Seasonal variation in the activity and dietary budgets of Cat Ba langurs (*Trachypithecus poliocephalus*). *International Journal of Primatology* 37:586–604.

Hennessy, S. M., Wisinski, C. L., Ronan, N. A., Gregory, C. J., Swaisgood, R. R., and Nordstrom, L. A. 2022. Release strategies and ecological factors influence mitigation translocation outcomes for burrowing owls: A comparative evaluation. *Animal Conservation* 25:614–626.

Henzi, S. P., Brown, L. R., Barrett, L., and Marais, A. J. 2011. Troop size, habitat use, and diet of chacma baboons (*Papio hamadryas ursinus*) in commercial pine plantations: Implications for management. *International Journal of Primatology* 32:1020–1032.

Herbinger, I., Boesch, C., and Rothe, H. 2001. Territory characteristics among three neighboring chimpanzee communities in the Taï National Park, Côte d'Ivoire. *International Journal of Primatology* 22:143–167.

Hilgartner, R., Fichtel, C., Kappeler, P. M., and Zinner, D. 2012. Determinants of pair-living in red-tailed sportive lemurs (*Lepilemur ruficaudatus*). *Ethology* 118:466–479.

Hill, R. A. 2006. Thermal constraints on activity scheduling and habitat choice in baboons. *American Journal of Physical Anthropology* 129:242–249.

Hill, R. A., and Dunbar, R. I. M. 1998. An evaluation of the roles of predation rate and predation risk as selective pressures on primate grouping behaviour. *Behaviour* 135:411–430.

Hoffman, T. S., and O'Riain, M. J. 2012. Troop size and human-modified habitat affect the ranging patterns of a chacma baboon population in the Cape Peninsula, South Africa. *American Journal of Primatology* 74:853–863.

Hoogland, J. L. 1979. The effect of colony size on individual alertness of prairie dogs (Sciuridae: *Cynomys* spp.). *Animal Behaviour* 27:394–407.

Hopkins, M. E. 2011. Mantled howler (*Alouatta palliata*) arboreal pathway networks: Relative impacts of resource availability and forest structure. *International Journal of Primatology* 32:238–258.

Homewood, K. M. 1978. Feeding strategy of the Tana mangabey (*Cercocebus galeritus galeritus*) (Mammalia: Primates). *Journal of Zoology, London* 186:375–391.

Hu, N., Guan, Z., Huang, B., Ning, W., He, K., Fan, P., and Jiang, X. 2018. Dispersal and female philopatry in a long-term, stable, polygynous gibbon population: Evidence from 16 years field observation and genetics. *American Journal of Primatology* 80:e22922. DOI: 10.1002/ajp.22922

Huang, C., Wei, F., Li, M., Li, Y., and Sun, R. 2003. Sleeping cave selection, activity pattern and time budget of white-headed langurs. *International Journal of Primatology* 24:813–824.

Huang, Z., Yuan, P., Huang, H., Tang, X., Xu, W., Huang, C., and Zhou, Q. 2017. Effect of habitat fragmentation on ranging behavior of white-headed langurs in limestone forests in Southwest China. *Primates* 58:423–434.

Huck, M., and Fernandez-Duque, E. 2017. The floaters dilemma: Use of space by wild Azara's owl monkeys, *Aotus azarae*, in relation to group ranges. *Animal Behaviour* 127:33–41.

Inoue, N., and Shimada, M. 2020. Comparisons of activity budgets, interactions, and social structures in captive and wild chimpanzees (*Pan troglodytes*). *Animals* 10:1063. DOI: 10.3390/ani10061063

Inoue, Y., Sinun, W., and Okanoya, K. 2016. Activity budget, travel distance, sleeping time, height of activity and travel order of wild East Bornean grey gibbons (*Hylobates funereus*) in Danum Valley Conservation Area. *Raffles Bulletin of Zoology* 64:127–138.

Isbell, L. A. 1983. Daily ranging behavior of red colobus (*Colobus badius tephrosceles*) in Kibale Forest, Uganda. *Folia Primatologica* 41:34–48.

Isbell, L. A. 1990. Sudden short-term increase in mortality of vervet monkeys (*Cercopithecus aethiops*) due to leopard predation in Amboseli National Park, Kenya. *American Journal of Primatology* 21:41–52.

Isbell, L. A. 1991. Contest and scramble competition: Patterns of female aggression and ranging behavior among primates. *Behavioral Ecology* 2:143–155.

Isbell, L. A. 1994. Predation on primates: Ecological patterns and evolutionary consequences. *Evolutionary Anthropology* 3:61–71.

Isbell, L. A. 1998. Diet for a small primate: Insectivory and gummivory in the (large) patas monkey (*Erythrocebus patas*). *American Journal of Primatology* 45:381–398.

Isbell, L. A. 2004. Is there no place like home? Ecological bases of female dispersal and philopatry and their consequences for the formation of kin groups. In *Kinship and Behavior in Primates* (Chapais, B., and Berman, C., eds.). Oxford University Press, New York, pp. 71–108.

Isbell, L. A. 2006. Snakes as agents of evolutionary change in primate brains. *Journal of Human Evolution* 51:1–35.

Isbell, L. A. 2009. *The Fruit, the Tree, and the Serpent: Why We See So Well*. Harvard University Press, New York.

Isbell, L. A. 2012. Re-evaluating the Ecological Constraints model with red colobus monkeys (*Procolobus rufomitratus tephrosceles*). *Behaviour* 149:493–529.

Isbell, L. A., and Bidner, L. R. 2016. Vervet monkey (*Chlorocebus pygerythrus*) alarm calls to leopards (*Panthera pardus*) function as a predator deterrent. *Behaviour* 153:591–606.

Isbell, L. A., Bidner, L. R, Loftus, J. C., Kimuyu, D., and Young, T. P. 2021. Absentee owners and overlapping home ranges in a territorial species. *Behavioral Ecology and Sociobiology* 75:21. DOI: 10.1007/s00265-020-02945-7

Isbell, L. A., Bidner, L. R., Van Cleave, E. K., Matusmoto-Oda, A., and Crofoot, M. C. 2018. GPS-identified vulnerabilities of savannah-woodland primates to leopard predation and their implications for early hominins. *Journal of Human Evolution* 118:1–13.

Isbell, L. A., Cheney, D. L., and Seyfarth, R. M. 1990. Costs and benefits of home range shifts among vervet monkeys (*Cercopithecus aethiops*) in Amboseli National Park, Kenya. *Behavioral Ecology and Sociobiology* 27:351–358.

Isbell, L. A., Cheney, D. L., and Seyfarth, R. M. 1991. Group fusions and minimum group sizes in vervet monkeys (*Cercopithecus aethiops*). *American Journal of Primatology* 25:57–65.

Isbell, L. A., Cheney, D. L., and Seyfarth, R. M. 1993. Are immigrant vervet monkeys, *Cercopithecus aethiops*, at greater risk of mortality than residents? *Animal Behaviour* 45:729–734.

Isbell, L. A., Cheney, D. L., and Seyfarth, R. M. 2002. Why vervet monkeys (*Cercopithecus aethiops*) live in multimale groups. In *The Guenons: Diversity and Adaptation in African Monkeys* (Glenn, M., and Cords, M., eds.). Kluwer Academic/Plenum Publishers, New York, pp. 173–187.

Isbell, L. A., and Etting, S. F. 2017. Scales drive visual detection, attention, and memory of snakes in wild vervet monkeys (*Chlorocebus pygerythrus*). *Primates* 58:121–129.

Isbell, L. A., and Jaffe, K. L. E. 2013. *Chlorocebus pygerythrus* Vervet monkey. In *The Mammals of Africa: Vol II Primates* (Butynski, T. M., Kingdon, J., and Kalina, J., eds.). Bloomsbury Publishing, London, pp. 277–283.

Isbell, L. A., Pruetz, J. D., and Young, T. P. 1998b. Movements of vervets (*Cercopithecus aethiops*) and patas monkeys (*Erythrocebus patas*) as estimators of food resource size, density, and distribution. *Behavioral Ecology and Sociobiology* 42:123–133.

Isbell, L. A., Pruetz, J. D., Lewis, M., and Young, T. P. 1998a. Locomotor activity differences between sympatric patas monkeys (*Erythrocebus patas*) and vervet monkeys (*Cercopithecus aethiops*): Implications for the evolution of long hindlimb length in *Homo*. *American Journal of Physical Anthropology* 105:199–207.

Isbell, L. A., and Van Vuren, D. 1996. Differential costs of locational and social dispersal and their consequences for female group-living primates. *Behaviour* 133:1–36.

Isbell, L. A., and Young, T. P. 1993a. Social and ecological influences on activity budgets of vervet monkeys, and their implications for group living. *Behavioral Ecology and Sociobiology* 32:377–385.

Isbell, L. A., and Young, T. P. 1993b. Human presence reduces predation in a free-ranging vervet monkey population in Kenya. *Animal Behaviour* 45:1233–1235.

Isbell, L. A., and Young, T. P. 1996. The evolution of bipedalism in hominids and reduced group size in chimpanzees: Alternative responses to decreasing resource availability. *Journal of Human Evolution* 30:389–397.

Isbell, L. A., Young, T. P., Jaffe, K. E., Carlson, A. A., and Chancellor, R. L. 2009. Demography and life histories of sympatric patas monkeys, *Erythrocebus patas*, and vervets, *Cercopithecus aethiops*, in Laikipia, Kenya. *International Journal of Primatology* 30:103–124.

Iwamoto, T., and Dunbar, R. I. 1983. Thermoregulation, habitat quality and the behavioural ecology of gelada baboons. *Journal of Animal Ecology* 52:357–366.

Izar, P., Verderane, M. P., Peternelli-Dos-Santos, L., Mendonça-Furtado, O., Presotto, A., Tokuda, M., Visalberghi, E., and Fragaszy, D. 2012. Flexible and conservative features of social systems of tufted capuchin monkeys: Comparing the socioecology of *Sapajus libidinosus* and *Sapajus nigritus*. *American Journal of Primatology* 74:315–331.

Jack, K. M., and Isbell, L. A. 2009. Dispersal in primates: Advancing an individualized approach. *Behaviour* 146:429–436.

Jacobs, G. H., Deegan, J. F., II, Neitz, J., Crognale, M. A., and Neitz, M. 1993. Photopigments and color vision in the nocturnal monkey, *Aotus*. *Vision Research* 33:1773–1783.

Jaffe, K. E., and Isbell, L. A. 2010. Changes in ranging and agonistic behavior of vervet monkeys (*Cercopithecus aethiops*) after predator-induced group fusion. *American Journal of Primatology* 72:634–644.

Jaman, M. F., and Huffman, M. A. 2013. The effect of urban and rural habitats and resource type on activity budgets of commensal rhesus macaques (*Macaca mulatta*) in Bangladesh. *Primates* 54:49–59.

Janmaat, K. R. L., Byrne, R. W., and Zuberbühler, K. 2006. Evidence for a spatial memory of fruiting states of rainforest trees in wild mangabeys. *Animal Behaviour* 72:797–807.

Janmaat, K. R. L., and Chancellor, R. L. 2010. Exploring new areas: How important is long-term spatial memory for mangabey (*Lophocebus albigena johnstonii*) foraging efficiency? *International Journal of Primatology* 31:863–886.

Janmaat, K. R. L., Olupot, W., Chancellor, R. L., Arlet, M. E., and Waser, P. M. 2009. Long-term site fidelity and individual home range shifts in *Lophocebus albigena*. *International Journal of Primatology* 30:443–466.

Janson, C. H. 1985. Aggressive competition and individual food consumption in wild brown capuchin monkeys (*Cebus apella*). *Behavioral Ecology and Sociobiology* 18:125–138.

Janson, C. H., and Di Bitetti, M. S. 1997. Experimental analysis of food detection in capuchin monkeys: Effects of distance, travel speed, and resource size. *Behavioral Ecology and Sociobiology* 41:17–24.

Janson, C. H., and Goldsmith, M. L. 1995. Predicting group size in primates: Foraging costs and predation risks. *Behavioral Ecology* 6:326–336.

Janson, C. H., and van Schaik, C. P. 1988. Recognizing the many faces of primate food competition: Methods. *Behaviour* 105:165–186.

Jarman, P. J. 1974. The social organization of antelope in relation to their ecology. *Behaviour* 48:215–267.

Jay, P. C. (ed). 1968. *Primates: Studies in Adaptation and Variability*. Holt, Rinehart, and Winston, New York.

Johnson, T. J. M. 1985. Lion-tailed macaque behavior in the wild. In *The Lion-Tailed Macaque: Status and Conservation* (Heltne, P. G., ed.). Alan R. Liss, New York, pp. 41–63.

Jolly, A. 1999. *Lucy's Legacy: Sex and Intelligence in Human Evolution*. Harvard University Press, Cambridge, MA.

Jolly, A., and Pride, E. 1991. Troop histories and range inertia of *Lemur catta* at Berenty, Madagascar: A 33-year perspective. *International Journal of Primatology* 20:359–373.

Joly, M., and Zimmermann, E. 2007. First evidence for relocation of stationary food resources during foraging in a strepsirhine primate (*Microcebus murinus*). *American Journal of Primatology* 69:1045–1052.

Joly, M., and Zimmermann, E. 2011. Do solitary foraging nocturnal mammals plan their routes? *Biology Letters* 7:638–640.

Jorde, L. B., and Spuhler, J. N. 1974. A statistical analysis of selected aspects of primate demography, ecology, and sexual behavior. *Journal of Anthropological Research* 30:199–224.

Jung, L., Mourthe, I., Grelle, C. E. V., Strier, K. B., and Boubli, J. P. 2015. Effects of local habitat variation on the behavioral ecology of two sympatric groups of brown howler monkeys (*Alouatta clamitans*). *PLoS ONE* 10:e0129789. DOI: 10.1371/journal.pone.0129789

Júnior, O. F., de Oliveira Porfirio, G. E., Santos, F. M., Nantes, W. A. G., de Assis, W. O., de Andrade, G. B., Herrera, H. M., and Rímoli, J. 2019. Behavioral activities and diet of Azaras's capuchin monkey, *Sapajus cay* (Illiger, 1815), in a forest remnant of the Brazilian Cerrado. *Studies on Neotropical Fauna and Environment* 55:149–154. DOI: 10.1080/01650521.2019.1708228

Justa, P., Kumar, R. S., Talukdar, G., and Sinha, A. 2019. Sharing from the same bowl: Resource partitioning between sympatric macaque species in the western Himalaya, India. *International Journal of Primatology* 40:356–373.

Kahlenberg, S. M., Emery Thompson, M., Muller, M. N., and Wrangham, R. W. 2008. Immigration costs for female chimpanzees and male protection as an immigrant counterstrategy to intrasexual aggression. *Animal Behaviour* 76:1497–1509.

Kanagasuntheram, R., and Krishnamurti, A. 1965. Observations on the carotid rete in the lesser bush baby (*Galago senegalensis senegalensis*). *Journal of Anatomy* 99:861–875.

Kanamori, T., Kuze, N., Bernard, H., Malim, T. P., and Kohshima, S. 2010. Feeding ecology of Bornean orangutans (*Pongo pygmaeus morio*) in Danum Valley, Sabah, Malaysia: A 3-year record including two mast fruitings. *American Journal of Primatology* 72:820–840.

Kane, E. E., and McGraw, W. S. 2018. Effects of chimpanzee (*Pan troglodytes*) hunting seasonality and red colobus (*Piliocolobus badius*) association on Diana monkeys (*Cercopithecus diana*) in Taï National Park, Côte d'Ivoire. *International Journal of Primatology* 39:532–546.

Kappeler, P. M. 2000. Causes and consequences of unusual sex ratios among lemurs. In *Primate Males: Causes and Consequences of Variation in Group Composition* (Kappeler, P. M., ed.). Cambridge University Press, Cambridge, UK, pp. 55–63.

Kappeler, P. M., and van Schaik, C. P. 2002. Evolution of primate social systems. *International Journal of Primatology* 23:707–740.

Kappeler, P. M., and Fichtel, C. 2012. Female reproductive competition in *Eulemur rufifrons*: Eviction and reproductive restraint in a plurally breeding Malagasy primate. *Molecular Ecology* 21:685–698.

Kappeler, P. M., and Fichtel, C. 2016. The evolution of *Eulemur* social organization. *International Journal of Primatology* 37:10–28.

Kappeler, P. M., Mass, V., and Port, M. 2009. Even adult sex ratios in lemurs: Potential costs and benefits of subordinate males in Verreaux's sifaka (*Propithecus verreauxi*) in the Kirindy Forest CFPF, Madagascar. *American Journal of Physical Anthropology* 140:487–497.

Kappeler, P. M., Pereira, M. E., and van Schaik, C. P. 2003. Primate life histories and socioecology. In *Primate Life Histories and Socioecology* (Kappeler, P. M., and Pereira, M. E., eds.). University of Chicago Press, Chicago, IL, pp. 1–23.

Karr, K. 1982. The ecology and behavior of the siamang (*Hylobates syndactylus*) in Sumatera. Unpublished Master's thesis, University of California, Davis.

Katsir, Z., and Crewe, R. M. 1980. Chemical communication in *Galago crassicaudatus*: investigation of the chest gland secretion. *South African Journal of Zoology* 15:249–254.

Kavana, T. S., Erinjery, J. J., and Singh, M. 2015. Folivory as a constraint on social behaviour of langurs in South India. *Folia Primatologica* 86:420–431.

Kawai, N. 2019. *The Fear of Snakes: Evolutionary and Psychobiological Perspectives on Our Innate Fear*. Springer Nature Singapore Pte Ltd., Singapore.

Kawai, N., and He, H. 2016. Breaking snake camouflage: Humans detect snakes more accurately than other animals under less discernible visual conditions. *PLoS ONE* 11: e0164342. DOI: 10.137/journal.pone.0164342

Khatun, M. T., Jaman, M. F., Rahman, M. M., and Alam, M. M. 2015. The effect of urban and rural habitats on activity budgets of the endangered Northern Plains sacred langur, *Semnopithecus entellus* (Dufresne, 1797) in Jessore, Bangladesh. *Mammalia* 82:423–430.

Kim, S., Lappan, S., and Choe, J. C. 2011. Diet and ranging behavior of the endangered Javan gibbon (*Hylobates moloch*) in a submontane tropical rainforest. *American Journal of Primatology* 73:270–280.

Kirchhof, J., and Hammerschmidt, K. 2006. Functionally referential alarm calls in tamarins (*Saguinus fuscicollis* and *Saguinus mystax*)—evidence from playback experiments. *Ethology* 112:346–354.

Knott, C., Beaudrot, L., Snaith, T., White, S., Tschauner, H., and Planansky, G. 2008. Female-female competition in Bornean orangutans. *International Journal of Primatology* 29:975–997.

Koda, H., Shimooka, Y., and Sugiura, H. 2008. Effects of caller activity and habitat visibility on contact call rate of wild Japanese macaques (*Macaca fuscata*). *American Journal of Primatology* 70:1055–1063.

Koenig, A. 2002. Competition for resources and its behavioral consequences among female primates. *International Journal of Primatology* 23:759–783.

Koenig, A., Beise, J., Chalise, M. K., and Ganzhorn, J. U. 1998. When females should contest for food—testing hypotheses about resource density, distribution, size, and quality with Hanuman langurs (*Presybytis entellus*). *Behavioral Ecology and Sociobiology* 42:225–237.

Koga, A., Tanabe, H., Hirai, Y., Imai, H., Imamura, M., Oishi, T., Stanyon, R., and Hirai, H. 2017. Co-opted megasatellite DNA drives evolution of secondary night vision in Azara's owl monkey. *Genome Biology and Evolution* 9:1963–1970.

Koirala, S., Chalise, M. K., Katuwal, H. B., Gaire, R., Pandey, B., and Ogawa, H. 2017. Diet and activity of *Macaca assamensis* in wild and semi-provisioned groups in Shivapuri Nagarjun National Park, Nepal. *Folia Primatologica* 88:57–74.

Komers, P. E., and Brotherton, P. N. M. 1997. Female space use is the best predictor of monogamy in mammals. *Proceedings of the Royal Society London B: Biological Sciences* 264:1261–1270.

Korstjens, A. H. 2001. *The Mob, the Secret Sorority, and the Phantoms: an Analysis of the Socio-Ecological Strategies of the Three Colobines of Taï.* PhD dissertation, Utrecht University, Utrecht, Netherlands.

Korstjens, A. H., Bergmann, K., Defferenez, C., Krebs, M., Nijssen, E. C., van Oirschot, B. A. M., Paukert, C., and Schippers, E. Ph. 2007. How small-scale differences in food competition lead to different social systems in three closely related sympatric colobines. In *Monkeys of the Taï Forest: an African Primate Community* (McGraw, W. S., Zuberbühler, K., and Noë, R., eds.). Cambridge University Press, Cambridge, UK, pp. 72–108.

Korstjens, A. H., Hillyer, A. P., and Koné, I. 2022. Red colobus natural history. In *The Colobines: Natural History, Behaviour and Ecological Diversity* (Matsuda, I., Grueter, C. C., and Teichroeb, J. A., eds.). Cambridge University Press, Cambridge, UK, pp. 108–127.

Korstjens, A. H., Nijssen, E. C., and Noë, R. 2005. Intergroup relationships in western black-and-white colobus, *Colobus polykomos polykomos*. *International Journal of Primatology* 26:1267–1289.

Korstjens, A. H., and Schippers, E. Ph. 2003. Dispersal patterns among olive colobus in Taï National Park. *International Journal of Primatology* 24:515–539.

Kulp, J., and Heymann EW. 2015. Ranging, activity budget, and diet composition of red titi monkeys (*Callicebus cupreus*) in primary forest and forest edge. *Primates* 56:273–278.

Kumar, R. S., Mishra, C., and Sinha, A. 2007. Foraging ecology and time-activity budget of the Arunachal macaque *Macaca munzala*—a preliminary study. *Current Science* 93:532–539.

Kummer, H. 1968. *Social Organization of Hamadryas Baboons.* University of Chicago Press, Chicago, IL.

Kunz, J. A., Duvot, G. J., van Noordwijk, M. A., Willems, E. P., Townsend, M., Mardianah, N., Atmoko, S. S. U., Vogel, E. R., Nugraha, T. P., Heistermann, M.,

Agil, M., Weingrill, T., and van Schaik, C. P. 2021. The cost of associating with males for Bornean and Sumatran female orangutans: a hidden form of sexual conflict? *Behavioral Ecology and Sociobiology* 75:6. DOI: 10.1007/s00265-020-02948-4

Kurup, G. U., and Kumar, A. 1993. Time budget and activity patterns of the lion-tailed macaque (*Macaca silenus*). *International Journal of Primatology* 14:27–39.

LaBarge, L. R., Allan, A. T. L., Berman, C. M., Hill, R. A., and Margulis, S. W. 2021. Extent of threat detection depends on predator type and behavioral context in wild samango monkey groups. *Behavioral Ecology and Sociobiology* 75:13. DOI: 10.1007/s00265-020-02959-1

Lappan, S. 2009. The effects of lactation and infant care on adult energy budgets in wild siamangs (*Symphalangus syndactylus*). *American Journal of Physical Anthropology* 140:290–301.

Lappan, S., Sibarani, M., O'Brien, T. G., Nurcahyo, A., Andayani, N., Rustiati, E. L., Surya, R. A., and Morino, L. 2021. Long-term effects of forest fire on habitat use by siamangs in southern Sumatra. *Animal Conservation* 24:355–366.

Lawes, M. J., and Piper, S. E. 1992. Activity patterns in free-ranging samango monkeys (*Cercopithecus mitis erythrarchus* Peters, 1852) at the southern range limit. *Folia Primatologica* 59:186–202.

Le, H. T., Hoang, D. M., and Covert, H. H. 2019. Diet of the Indochinese silvered langur (*Trachypithecus germaini*) in Kien Luong Karst area, Kien Giang Province. *American Journal of Primatology* 81:e23041. DOI: 10.1002/ajp.23041

Le, Q. V., Isbell, L. A., Matsumoto, J., Quang, L. V., Hori, E., Tran, A. H., Maior, R. S., Tomaz, C., Ono, T., and Nishijo, H. 2014. Monkey pulvinar neurons fire differentially to snake postures. *PLoS ONE* 9:e114258. DOI: 10.1371/journal.pone.0114258

Le, Q. V., Isbell, L. A., Nguyen, M. N., Matsumoto, J., Hori, E., Maior, R. S., Tomaz, C., Tran, A. H., Ono, T., and Nishijo, H. 2013. Pulvinar neurons reveal neurobiological evidence of past selection for rapid detection of snakes. *Proceedings of the National Academy of Sciences USA* 110:19000–19005.

Lehman, S. M., Prince, W., and Mayor, M. 2001. Variations in group size in white-faced sakis (*Pithecia pithecia*): Evidence for monogamy or seasonal congregations? *Neotropical Primates* 9:96–101.

Lehmann, J., and Boesch, C. 2005. Bisexually bonded ranging in chimpanzees (*Pan troglodytes verus*). *Behavioral Ecology and Sociobiology* 57:525–535.

Le Malo, Y., Goffart, M., Rochas, A., Felbabel, H., and Chatonnet, J. 1981. Thermoregulation in the only nocturnal simian: The night monkey *Aotus trivirgatus*.

American Journal of Physiology—Regulatory, Integrative and Comparative Physiology 240:R156–R165.

Lemke, T. O. 1984. Foraging ecology of the long-nosed bat, *Glossophaga soricine*, with respect to resource availability. *Ecology* 65:538–548.

Lemoine, S., Preis, A., Samuni, L., Boesch, C., Crockford, C., and Wittig, R. M. 2020. Between-group competition impacts reproductive success in wild chimpanzees. *Current Biology* 30:312–318.

León, J., Thiriau, C., Crockford, C., and Zuberbühler, K. 2023. Comprehension of own and other species' alarm calls in sooty mangabey vocal development. *Behavioral Ecology and Sociobiology* 77:56. DOI: 10.1007/x00265-023-03318–6

Levy, O., Dayan, T., Porter, W. P., and Kronfeld-Schor, N. 2019. Time and ecological resilience: Can diurnal animals compensate for climate change by shifting to nocturnal activity? *Ecological Monographs* 89:e01334. DOI: 10.1002/ecm.1334

Lewis, R. J. 2008. Social influences on group membership in *Propithecus verreauxi verreauxi*. *International Journal of Primatology* 29:1249–1270.

Li, Y., Li, D., Ren, B., Hu, J., Li, B., Krzton, A., and Li, M. 2014. Differences in the activity budgets of Yunnan snub-nosed monkeys (*Rhinopithecus bieti*) by age-sex class at Xiangguqing in Baimaxueshan Nature Reserve, China. *Folia Primatologica* 85:335–342.

Li, Y., Ma, G., Zhou, Q., and Huang, Z. 2020. Seasonal variation in activity budget of Assamese macaques in limestone forest of Southwest Guangxi, China. *Folia Primatologica* 91:495–511.

Li, Z., and Rogers, E. 2004. Habitat quality and activity budgets of white-headed langurs in Fusui, China. *International Journal of Primatology* 25:41–54.

Lima, M., Mendes, S. L., and Strier, K. B. 2019. Habitat use in a population of the northern muriqui (*Brachyteles hypoxanthus*). *International Journal of Primatology* 40:470–495.

Linn, I., and Key, G. 1996. Use of space by the African Striped Ground Squirrel *Xerus erythropus*. *Mammal Review* 26:9–26.

Liu, J., and Bhumpakphan, N. 2020. Comparison of activity budgets, diet, and habitat utilization between provisioned and wild groups of the François' langur (*Trachypithecus francoisi*) in Mayanghe National Nature Reserve, China. *Folia Primatologica* 91:15–20.

Liu, X., Stanford, C. B., and Li, Y. 2013. Effect of group size on time budgets of Sichuan snub-nosed monkeys (*Rhinopithecus roxellana*) in Shennongjia National Nature Reserve, China. *International Journal of Primatology* 34:349–360.

Lopes, K. G. D., and Bicca-Marques, J. C. 2017. Ambient temperature and humidity modulate the behavioural thermoregulation of a small arboreal mammal (*Callicebus bernhardi*). *Journal of Thermal Biology* 69:104–109.

Löttker, P., Huck, M., and Heymann, E. W. 2004. Demographic parameters and events in wild moustached tamarins (*Saguius mystax*). *American Journal of Primatology* 64:425–449.

Lovegrove, B. G., Canale, C., Levesque, D., Fluch, G., Řeháková-Petrů, M., and Ruf, T. 2014. Are tropical small mammals physiologically vulnerable to Arrhenius effects and climate change? *Physiological and Biochemical Zoology* 87:30–45.

Lucchesi, S., Cheng, L., Janmaat, K., Mundry, R., Pisor, A., and Surbeck, M. 2020. Beyond the group: How food, mates, and group size influence intergroup encounters in wild bonobos. *Behavioral Ecology* 31:519–532.

Lührs, M.-L., Dammhahn, M., Kappeler, P. M., and Fichtel, C. 2009. Spatial memory in the grey mouse lemur (*Microcebus murinus*). *Animal Cognition* 12:599–609.

Lukas, D., and Clutton-Brock, T. H. 2013. The evolution of social monogamy in mammals. *Science* 341:526–530.

Lukas, D., and Huchard, E. 2014. The evolution of infanticide by males in mammalian societies. *Science* 346:841–844.

MacDonald, S. E., and Agnes, M. M. 1999. Orangutan (*Pongo pygmaeus abelii*) spatial memory and behavior in a foraging task. *Journal of Comparative Psychology* 113:213–217.

MacKinnon, J. 1974. The behaviour and ecology of wild orang-utans (*Pongo pygmaeus*). *Animal Behaviour* 22:3–74.

Magliocca, F., and Gautier-Hion, A. 2002. Mineral content as a basis for food selection by western lowland gorillas in a forest clearing. *American Journal of Primatology* 57:67–77.

Maloney, S. K., Mitchell, D., Mitchell, G., and Fuller, A. 2007. Absence of selective brain cooling in unrestrained baboons exposed to heat. *American Journal of Physiology—Regulatory, Integrative and Comparative Physiology* 292:R2059–R2067.

Mamo, M., and Wube, T. 2019. Variability in group size and daily activity budget of family groups of the gelada baboon (*Theropithecus gelada*) at Guassa Community Conservation Area, Central Ethiopia. *Journal of Ecology and Environment* 43:26. DOI: 10.1186/s41610-019-0124-5

Mandl, I., Holderied, M., and Schwitzer, C. 2018. The effects of climate seasonality on behavior and sleeping site choice in Sahamalaza sportive lemurs, *Lepilemur sahamalaza*. *International Journal of Primatology* 39:1039–1067.

Marais, A. J., Brown, L. R., Barrett, L., and Henzi, S. P. 2006. Population structure and habitat use of baboons (*Papio hamadryas ursinus*) in the Blyde Canyon Nature Reserve. *Koedoe* 49:67–76.

Maré, C., Landman, M., and Kerley, G. I. H. 2019. Rocking the landscape: Chacma baboons (*Papio ursinus*) as zoogeomorphic agents. *Geomorphology* 327:504–510.

Maré, C., Landman, M., and Kerley, G. I. H. 2021. How should a clever baboon choose and move rocks? *Biotropica* 53:162–169.

Marsh, C. 1978. Comparative activity budgets of red colobus. In *Recent Advances in Primatology, Vol 1: Behaviour* (Chivers, D. J., and Herbert, J., eds.). Academic Press, New York, pp. 249–251.

Marsh, C. W. 1979. Female transference and mate choice among Tana River red colobus. *Nature* 281:568–569.

Masi, S., Cipolletta, C., and Robbins, M. M. 2009. Western lowland gorillas (*Gorilla gorilla gorilla*) change their activity patterns in response to frugivory. *American Journal of Primatology* 71:91–100.

Matsuda, I., Tuuga, A., and Higashi, S. 2009. The feeding ecology and activity budget of proboscis monkeys. *American Journal of Primatology* 71:478–492.

Matsudaira, K., Ishida, T., Malaivijitnond, S., and Reichard, U. H. 2017. Short dispersal distance of males in a wild white-handed gibbon (*Hylobates lar*) population. *American Journal of Physical Anthropology* 167:61–71.

Matsumoto-Oda, A., and Oda, R. 1998. Changes in the activity budget of cycling female chimpanzees. *American Journal of Primatology* 46:157–166.

Matthews, L. J. 2009. Activity patterns, home range size, and intergroup encounters in *Cebus albifrons* support existing models of capuchin socioecology. *International Journal of Primatology* 30:709–728.

McCain, C. M., and King, S. R. B. 2014. Body size and activity times mediate mammalian responses to climate change. *Global Change Biology* 20:1760–1769.

McFarland, R., Barrett, L., Costello, M.-A., Fuller, A., Hetem, R. S., Maloney, S. K., Mitchell, D., and Henzi, P. S. 2020. Keeping cool in the heat: Behavioral thermoregulation and body temperature patterns in wild vervet monkeys. *American Journal of Physical Anthropology* 171:407–418.

McGraw, W. S. 1998. Comparative locomotion and habitat use of six monkeys in the Tai Forest, Ivory Coast. *American Journal of Physical Anthropology* 105:493–510.

McKey, D., and Waterman, P. G. 1982. Ranging behaviour of a group of black colobus (*Colobus satanas*) in the Douala-Edea Reserve, Cameroon. *Folia Primatologica* 39:264–304.

McLean, K. A., Trainor, A. M., Asner, G. P., Crofoot, M. C., Hopkins, M. E., Campbell, C. J., Martin, R. E., Knapp, D. E., and Jansen, P. A. 2016. Movement patterns of three arboreal primates in a Neotropical moist forest explained by LiDAR-estimated canopy structure. *Landscape Ecology* 31:1849–1862.

McNab, B. K. 1963. Bioenergetics and the determination of home range size. *American Naturalist* 97:133–140.

McNab, B. K., and Wright, P. C. 1987. Temperature regulation and oxygen consumption in the Philippine tarsier *Tarsius syrichta*. *Physiological Zoology* 60:596–600.

McPherson, S. C., Brown, M., and Downs, C. T. 2016. Diet of the crowned eagle (*Stephanoaetus coronatus*) in an urban landscape: Potential for human-wildlife conflict? *Urban Ecosystems* 19:383–396.

Mekonnen, A., Bekele, A., Fashing, P. J., Hemson, G., and Atickem, A. 2010. Diet, activity patterns, and ranging ecology of the Bale monkey (*Chlorocebus djamdjamensis*) in Odobullu Forest, Ethiopia. *International Journal of Primatology* 31:339–362.

Mekonnen, A., Fashing, P. J., Bekele, A., Hernandez-Aguilar, R. A., Rueness, E. K., Nguyen, N., and Stenseth, N. C. 2017. Impacts of habitat loss and fragmentation on the activity budget, ranging ecology and habitat use of Bale monkeys (*Chlorocebus djamdjamensis*) in the southern Ethiopian highlands. *American Journal of Primatology* 79:e22644. DOI: 10.1002/ajp.22644

Ménard, N. 2004. Do ecological factors explain variation in social organization? In *Macaque Societies: A Model for the Study of Social Organization* (Thierry, B., Singh, M., and Kaumanns, W., eds.). Cambridge University Press, New York, pp. 237–262.

Ménard, N., and Vallet, D. 1997. Behavioral responses of Barbary macaques (*Macaca sylvanus*) to variations in environmental conditions in Algeria. *American Journal of Primatology* 43:285–304.

Meno, W., Coss, R. G., and Perry, S. 2013. Development of snake-directed antipredator behavior by wild white-faced capuchin monkeys: 1. Snake-species discrimination. *American Journal of Primatology* 75:281–291.

Menon, S., and Poirier, F. E. 1996. Lion-tailed macaques (*Macaca silenus*) in a disturbed forest fragment: activity patterns and time budget. *International Journal of Primatology* 17:969–985.

Mertl-Millhollen, A. S. 1988. Olfactory demarcation of territorial but not home range boundaries by *Lemur catta*. *Folia Primatologica* 50:175–187.

Mielke, A., Crockford, C., and Wittig, R. M. 2019. Snake alarm calls as a public good in sooty mangabeys. *Animal Behaviour* 158:201–209.

Mikula, P., Šaffa, G., Nelson, E., and Tryjanowski, P. 2018. Risk perception of vervet monkeys *Chlorocebus pygerythrus* to humans in urban and rural environments. *Behavioral Processes* 147:21–27.

Miller, L. E., and Treves, A. 2007. Predation on primates: Past studies, current challenges, and directions for the future. In *Primates in Perspective* (Campbell, C. J., Fuentes, A., MacKinnon, K. C., Panger, M., and Bearder, S. K., eds.). Oxford University Press, New York, pp. 525–543.

Milton, K. 1980. *The Foraging Strategy of Howler Monkeys*. Columbia University Press, New York.

Milton, K. 1984. Habitat, diet, and activity patterns of free-ranging woolly spider monkeys (*Bracheteles arachnoides* E. Geoffroy 1806). *International Journal of Primatology* 5:491–514.

Milton, K., and May, M. L. 1976. Body weight, diet and home range area in primates. *Nature* 259:459–462.

Mir, Z. R., Noor, A., Habib, B., and Veeraswami, G. G. 2015. Seasonal population density and winter survival strategies of endangered Kashmir gray langur (*Semnopithecus ajax*) in Dachigam National Park, Kashmir, India. *SpringerPlus* 4:562. DOI: 10.1186/s40064-015-1366-z

Mitani, J. C. 1989. Orangutan activity budgets: Monthly variations and the effects of body size, parturition, and sociality. *American Journal of Primatology* 18:87–100.

Mitani, J. C. 1990. Demography of agile gibbons. *International Journal of Primatology* 11:441–424.

Mitani, J. C., Grether, G. F., Rodman, P. S., and Priatna, D. 1991. Associations among wild orang-utans: Sociality, passive aggregations or chance? *Animal Behaviour* 42:33–46.

Mitani, J. C., Gros-Louis, J., and Manson, J. H. 1996. Number of males in primate groups: Comparative tests of competing hypotheses. *American Journal of Primatology* 3:315–332.

Mitani, J. C., Watts, D. P., and Amsler, S. J. 2010. Lethal intergroup aggression leads to territorial expansion in wild chimpanzees. *Current Biology* 20:R507–R508. DOI: 10.1016/j.cub.2010.04.021

Miyamoto, M. M., Allen, J. M., Gogarten, J. F., and Chapman, C. A. 2013. Microsatellite DNA suggests that group size affects sex-biased dispersal patterns in red colobus monkeys. *American Journal of Primatology* 75:478–490.

Montgomery, S. H., and Mundy, N. I. 2013. Parallel episodes of phyletic dwarfism in callitrichid and cheirogaleid primates. *Journal of Evolutionary Biology* 26:810–819.

Mörchen, J., Luhn, F., Wassmer, O., Kunz, J. A., Kulik, L., van Noordwijk, M. A., van Schaik, C. P., Rianti, P., Utami Atmoko, S. S., Widdig, A., and Schuppli, C. 2023. Migrant orangutan males use social learning to adapt to new habitat after dispersal. *Frontiers in Ecology and Evolution* 11:1158887. DOI: 10.3389/fevo.2023.1158887

Morelli, T. L., King, S. J., Pochron, S. T., and Wright, P. C. 2009. The rules of engagement: Takeovers, infanticide and dispersal in a rainforest lemur, *Propithecus edwardsi*. *Behaviour* 146:499–523.

Morrogh-Bernard, H. C., Husson, S. J., Knott, C. D., Wich, S. A., van Schaik, C. P., van Noordwijk, M. A., Lackman-Ancrenaz, I., Marshall, A. J., Kanamori, T., Kuze, N., and bin Sakong, R. 2009. Orangutan activity budgets and diet: A comparison between species, populations and habitats. In *Orangutans: Geographic Variation in Behavioral Ecology and Conservation* (Wich, S. A., Utami Atmoko, S. S., Mitra Setia, T., and van Schaik, C. P., eds.). Oxford University Press, Oxford, pp. 119–133.

Müller, A. E. 1998. A Preliminary report on the social organization of *Cheirogaleus medius* (Cheirogaleidae; Primates) in North-West Madagascar. *Folia Primatologica* 69:160–166.

Müller, A. E., and Thalmann, U. 2000. Origin and evolution of primate social organisation: A reconstruction. *Biological Reviews* 75:405–435.

Murray, C. M., Mane, S. V., and Pusey, A. E. 2007. Dominance rank influences female space use in wild chimpanzees, *Pan troglodytes*: Towards an ideal despotic distribution. *Animal Behaviour* 74:1795–1804.

Nagel, U. 1973. A comparison of Anubis baboons, Hamadryas baboons and their hybrids at a species border in Ethiopia. *Folia Primatologica* 19:104–165.

Nakagawa, N. 2000. Seasonal, sex, and interspecific differences in activity time budgets and diets of patas monkeys (*Erythrocebus patas*) and tantalus monkeys (*Cercopithecus aethiops tantalus*), living sympatrically in northern Cameroon. *Primates* 41:161–174.

Nash, L. T. 1976. Troop fission in free ranging baboons in the Gombe Stream National Park, Tanzania. *American Journal of Physical Anthropology* 44:63–78.

Nash, L. T. 1998. Vertical clingers and sleepers: Seasonal influences on the activities and substrate use of *Lepilemur leucopus* at Beza Mahafaly Special Reserve, Madagascar. *Folia Primatologica* 69:204–217.

Nash, L. T., and Harcourt, C. S. 1986. Social organization of galagos in Kenyan coastal forests: I. *Galago zanzibaricus. American Journal of Primatology* 10:339–355.

Naude, V. N., Smyth, L. K., Weideman, E. A., Krochuk, B. A., and Amar, A. 2019. Using web-sourced photography to explore the diet of a declining African raptor, the Martial Eagle (*Polemaetus bellicosus*). *Condor* 121:1–9.

Naya, D. E., Naya, H., and White, C. R. 2018. On the interplay among ambient temperature, basal metabolic rate, and body mass. *American Naturalist* 192:518–524.

Nekaris, K. A. I. 2001. Activity budget and positional behavior of the Mysore slender loris (*Loris tardigradus lydekkerianus*): Implications for slow climbing locomotion. *Folia Primatologica* 72:228–241.

Nekaris, K. A. I. 2003. Spacing system of the Mysore slender loris (*Loris lydekkerianus lydekkerianus*). *American Journal of Physical Anthropology* 121:86–96.

Nekaris, K. A. I. 2006. Social lives of adult Mysore slender lorises (*Loris lydekkerianus lydekkerianus*). *American Journal of Primatology* 68:171–1182.

Ni, Q., Xie, M., Grueter, C. C., Jiang, X., Xu, H., Yao, Y., Zhang, M., Li, Y., and Yang, J. 2015. Effects of food availability and climate on activity patterns of western black-crested gibbons in an isolated forest fragment in southern Yunnan, China. *Primates* 56:351–363.

Ning, W.-H., Guan, Z.-H., Huang, B., Fan, P.-F., and Jian, X.-L. 2019. Influence of food availability and climate on behavior patterns of western black-crested gibbons (*Nomascas concolor*) at Mt. Wuliang, Yunnan, China. *American Journal of Primatology* 81:e23068. DOI: 10.1002/ajp.23068

Nishida, T., Corp, N., Hamai, M., Hasegawa, T., Hiraiwa-Hasegawa, M., Hosaka, K., Hunt, K. D., Itoh, N., Kawanaka, K., Matsumoto-Oda, A., Mitani, J. C., Nakamura, M., Norikoshi, K., Sakamaki, T., Turner, L., Uehara, S., and Zamma, K. 2003. Demography, female life history, and reproductive profiles among the chimpanzees of Mahale. *American Journal of Primatology* 59:99–121.

Norconk, M. A. 2006. Long-term study of group dynamics and female reproduction in Venezuelan *Pithecia pithecia. International Journal of Primatology* 27:653–674.

Norconk, M. A. 2011. Sakis, uakaris, and titi monkeys: Behavioral diversity in a radiation of primate seed predators. In *Primates in Perspective, 2nd edition* (Campbell, C. J., Fuentes, A., MacKinnon, K. C., Bearder, S. K., and Stumpf, R. M., eds.). Oxford University Press, New York, pp. 122–139.

Norscia, I., Ramanamanjato, J. B., and Ganzhorn, J. U. 2012. Feeding patterns and dietary profile of nocturnal southern woolly lemurs (*Avahi meridionalis*) in Southeast Madagascar. *International Journal of Primatology* 33:150–167.

Noser, R., and Byrne, R. W. 2007. Travel routes and planning of visits to out-of-sight resources in wild chacma baboons, *Papio ursinus*. *Animal Behaviour* 73:257–266.

Noser, R., and Byrne, R. W. 2010. How do wild baboons (*Papio ursinus*) plan their routes? Travel among multiple high-quality food sources with inter-group competition. *Animal Cognition* 13:145–155.

Noser, R., and Byrne, R. W. 2014. Change point analysis of travel routes reveals novel insights into foraging strategies and cognitive maps of wild baboons. *American Journal of Primatology* 76:399–409.

Notman, H., and Rendall, D. 2005. Contextual variation in chimpanzee pant hoots and its implications for referential communication. *Animal Behaviour* 70:177–190.

Nunes, A. 1995. Foraging and ranging patterns in white-bellied spider monkeys. *Folia Primatologica* 65:85–99.

Oates, J. F. 1977a. The guereza and its food. In *Primate Ecology: Studies of Feeding and Ranging Behavior in Lemurs, Monkeys and Apes* (Clutton-Brock, T. H., ed.). Academic Press, New York, pp. 275–321.

Oates, J. F. 1977b. The social life of a black and white colobus monkey, *Colobus guereza*. *Zeitschrift für Tierpsychologie* 45:1–60.

O'Brien, T. G., and Kinnaird, M. F. 1997. Behavior, diet, and movements of the Sulawesi crested black macaque (*Macaca nigra*). *International Journal of Primatology* 18:321–351.

Ohashi, K., and Thomson, J. D. 2009. Trapline foraging by pollinators: Its ontogeny, economics and possible consequences for plants. *Annals of Botany* 103:1365–1378.

Olupot, W., Chapman, C. A., Brown, C. H., and Waser, P. M. 1994. Mangabey (*Cercocebus albigena*) population density, group size, and ranging: A twenty-year comparison. *American Journal of Primatology* 32:197–205.

Olupot, W., and Waser, P. M. 2001. Activity patterns, habitat use and mortality risks of mangabey males living outside social groups. *Animal Behaviour* 61:1227–1235.

Olupot, W., and Waser, P. M. 2005. Patterns of male residency and intergroup transfer in gray-cheeked mangabeys (*Lophocebus albigena*). *American Journal of Primatology* 66:331–349.

Onderdonk, D. A., and Chapman, C. A. 2000. Coping with forest fragmentation: The primates of Kibale National Park, Uganda. *International Journal of Primatology* 21:587–611.

Opie, C., Atkinson, Q. D., Dunbar, R. I. M., and Shultz, S. 2013. Male infanticide leads to social monogamy in primates. *Proceedings of the National Academy of Sciences USA* 110:13328–13332.

Ostner, J., and Kappeler, P. M. 2004. Male life history and the unusual adult sex ratios of redfronted lemur, *Eulemur fulvus rufus*, groups. *Animal Behaviour* 67:249–259.

Ostro, L. E. T., Silver, S. C., Koontz, F. W., Horwich, R. H., and Brockett, R. 2001. Shifts in social structure of black howler (*Alouatta pigra*) groups associated with natural and experimental variation in population density. *International Journal of Primatology* 22:733–748.

Ostrofsky, K. R., and Robbins, M. M. 2020. Fruit-feeding and activity patterns of mountain gorillas (*Gorilla beringei beringei*) in Bwindi Impenetrable National Park, Uganda. *American Journal of Physical Anthropology* 173:3–20.

Overdorff, D. J. 1996. Ecological correlates to activity and habitat use of two prosimian primates: *Eulemur rubriventer* and *Eulemur fulvus rufus* in Madagascar. *American Journal of Primatology* 40:327–342.

Overdorff, D., Strait, S. G., and Telo, A. 1997. Seasonal variation in activity and diet in a small-bodied folivorous primate, *Hapalemur griseus*, in southeastern Madagascar. *American Journal of Primatology* 43:211223.

Owen-Smith, N. 2003. Foraging behavior, habitat suitability, and translocation success, with special reference to large mammalian herbivores. In *Animal Behavior and Wildlife Conservation* (Festa-Bianchet, M., and Apollonio, M., eds.). Island Press, Washington, DC, pp. 93–109.

Owen-Smith, N., and Mason, D. R. 2005. Comparative changes in adult and juvenile survival affecting population trends of African ungulates. *Journal of Animal Ecology* 74:762–773.

Packer, C. 1983. Demographic changes in a colony of Nile grassrats (*Arvicanthis niloticus*) in Tanzania. *Journal of Mammalogy* 64:159–161.

Palma, A. C., Vélez, A., Gómez-Posada, C., López, H., Zarate, D. A., and Stevenson, P. R. 2011. Use of space, activity patterns, and foraging behavior of red howler monkeys (*Alouatta seniculus*) in an Andean forest fragment in Colombia. *American Journal of Primatology* 73:1062–1071.

Palminteri, S., and Peres, C. A. 2012. Habitat selection and use of space by bald-faced sakis (*Pithecia irrorata*) in southwestern Amazonia: Lessons from a multiyear, multigroup study. *International Journal of Primatology* 33:401–417.

Palombit, R. 1994. Dynamic pair bonds in hylobatids: Implications regarding monogamous social systems. *Behaviour* 128:65–101.

Passamani, M. 1998. Activity budget of Geoffroy's marmoset (*Callithrix geoffroyi*) in an Atlantic forest in southeastern Brazil. *American Journal of Primatology* 46:333–340.

Pavelka, M. S. M., and Knopff, K. H. 2004. Diet and activity in black howler monkeys (*Alouatta pigra*) in southern Belize: Does degree of frugivory influence activity level? *Primates* 45:105–111.

Pearson, H. C. 2011. Sociability of female bottlenose dolphins (*Tursiops* spp.) and chimpanzees (*Pan troglodytes*): Understanding evolutionary pathways toward social convergence. *Evolutionary Anthropology* 20:85–95.

Pengfei, F., Garber, P., Chi, M., Guopeng, R., Changming, L., Xiaoyong, C., and Junxing, Y. 2015. High dietary diversity supports large group size in Indo-Chinese gray langurs in Wuliangshan, Yunnan, China. *American Journal of Primatology* 77:479–491.

Peres, C. A. 1989. Costs and benefits of territorial defense in wild golden lion tamarins, *Leontopithecus rosalia*. *Behavioral Ecology and Sociobiology* 25:227–233.

Phillips, P. K., and Heath, J. E. 1995. Dependency of surface temperature regulation on body size in terrestrial mammals. *Journal of Thermal Biology* 20:281–289.

Pichon, C., and Simmen, B. 2015. Energy management in crowned sifakas (*Propithecus coronatus*) and the timing of reproduction in a seasonal environment. *American Journal of Physical Anthropology* 158:269–278.

Pines, M., and Swedell, L. 2011. Not without a fair fight: Failed abductions of females in wild hamadryas baboons. *Primates* 52:249–252.

Pinheiro, T., Ferrari, S. F., and Lopes, M. A. 2013. Activity budget, diet, and use of space by two groups of squirrel monkeys (*Saimiri sciureus*) in eastern Amazonia. *Primates* 54:301–308.

Pinter-Wollman, N., Isbell, L. A., and Hart, L. A. 2009. Assessing translocation outcome: Comparing behavioral and physiological aspects of translocated and resident African elephants (*Loxodonta africana*). *Biological Conservation* 142:1116–1124.

Planck, M. 1949. *Scientific Autobiography and Other Papers* (Gaynor, F., transl). Philosophical Library, New York. Accessed from Open Road Integrated Media, Inc. ProQuest Ebook Central, 2014. http://ebookcentral.proquest.com/lib/ucdavis/detail.action?docID=1790404.

Pochron, S. T., and Wright, P. C. 2003. Variability in adult group compositions of a prosimian primate. *Behavioral Ecology and Sociobiology* 54:285–293.

Pope, T. R. 1989. *The Influence of Mating System and Dispersal Patterns on the Genetic Structure of Red Howler Monkey Populations*. PhD dissertation, University of Florida, Gainesville.

Pope, T. R. 2000. The evolution of male philopatry in neotropical monkeys. In *Primate Males: Causes and Consequences of Variation in Group Composition* (Kappeler, P. M., ed.). Cambridge University Press, Cambridge, UK, pp. 219–235.

Port, M., Johnstone, R. A., and Kappeler, P. M. 2010. Costs and benefits of multi-male associations in redfronted lemurs (*Eulemur fulvus rufus*). *Biology Letters* 6:620–622.

Porter, L. M. 2004. Forest use and activity patterns of *Callimico goeldii* in comparison to two sympatric tamarins, *Saguinus fuscicollis* and *Saguinus labiatus*. *American Journal of Physical Anthropology* 124:139–153.

Porter, L. M., and Garber, P. A. 2004. Goeldi's monkeys: A primate paradox? *Evolutionary Anthropology* 13:104–115.

Porter, L. M., and Garber, P. A. 2013. Foraging and spatial memory in wild Weddell's saddleback tamarins (*Saguinus fuscicollis weddelli*) when moving between distant and out-of-sight goals. *International Journal of Primatology* 34:30–48.

Porter, L. M., Sterr, S. M., and Garber, P. A. 2007. Habitat use and ranging behavior of *Callimico goeldii*. *International Journal of Primatology* 28:1035–1058.

Potts, K. B., Watts, D. P., and Wrangham, R. W. 2011. Comparative feeding ecology of two communities of chimpanzees (*Pan troglodytes*) in Kibale National Park, Uganda. *International Journal of Primatology* 32:669–690.

Poulsen, J. R., Jr., Clark, C. J., and Smith, T. B. 2001. Seasonal variation in the feeding ecology of the grey-cheeked mangabey (*Lophocebus albigena*) in Cameroon. *American Journal of Primatology* 54:91–105.

Powell, G. V. N. 1974. Experimental analysis of the social value of flocking by starlings (*Sturnus vulgaris*) in relation to predation and foraging. *Animal Behaviour* 22:501–505.

Power, M. L. 2010. Nutritional and digestive challenges to being a gum-feeding primate. In *The Evolution of Exudativory in Primates* (Burroughs, A. M., and Nash, L. T., eds.). Springer, New York, pp. 25–44.

Pozzi, L., Disotell, T. R., Bearder, S. K., Karlsson, J., Perkin, A., and Gamba, M. 2019. Species boundaries within morphologically cryptic galagos: Evidence from acoustic and genetic data. *Folia Primatologica* 90:279–299.

Printes, R. C., and Strier, K. B. 1999. Behavioral correlates of dispersal in female muriquis (*Brachyteles arachnoides*). *International Journal of Primatology* 20:941–960.

Promislow, D. E. L., and Harvey, P. H. 1990. Living fast and dying young: A comparative analysis of life-history variation among mammals. *Journal of Zoology, London* 220:417–437.

Prox, L., Fichtel, C., and Kappeler, P. M. 2023. Drivers and consequences of female reproductive competition in an egalitarian, sexually monomorphic primate. *Behavioral Ecology and Sociobiology* 77:53. DOI: 10.1007/s00265-023-03330-w

Pruetz, J. P. 2007. Evidence of cave use by savanna chimpanzees (*Pan troglodytes verus*) at Fongoli, Senegal: Implications for thermoregulatory behavior. *Primates* 48:316–319.

Pulliam, R. 1973. On the advantages of flocking. *Journal of Theoretical Biology* 38:419–422.

Pusey, A., Murray, C., Wallauer, W., Wilson, M., Wroblewski, E., and Goodall, J. 2008. Severe aggression among female *Pan troglodytes schweinfurthii* at Gombe National Park, Tanzania. *International Journal of Primatology* 29:949–973.

Pusey, A. E., and Packer, C. 1987. Dispersal and philopatry. In *Primate Societies* (Smuts, B. B., Cheney, D. L., Seyfarth, R. M., Wrangham, R. W., and Struhsaker, T. T., eds.). University of Chicago Press, Chicago, IL, pp. 250–266.

Pusey, A. E., and Schroepfer-Walker, K. 2013. Female competition in chimpanzees. *Philosophical Transactions of the Royal Society B* 368:201330077. DOI: 10.1098/rstb.2013.0077

Raboy, B. E., and Dietz, J. M. 2004. Diet, foraging, and use of space in wild golden-headed lion tamarins. *American Journal of Primatology* 63:1–15.

Radespiel, U. 2000. Sociality in the gray mouse lemur (*Microcebus murinus*) in north-western Madagascar. *American Journal of Primatology* 51:21–40.

Radespiel, U., Lutermann, H., Schmelting, B., Bruford, M. W., and Zimmerman, E. 2003. Patterns and dynamics of sex-biased dispersal in a nocturnal primate, the grey mouse lemur, *Microcebus murinus*. *Animal Behaviour* 65:709–719.

Radespiel, U., Reimann, W., Rahelinirina, M., and Zimmermann, E. 2006. Feeding ecology of sympatric mouse lemur species in northwestern Madagascar. *International Journal of Primatology* 27:311–321.

Radespiel, U., Sarikaya, Z., Zimmermann, E., and Bruford, M. W. 2001. Sociogenetic structure in a free-living nocturnal primate population: Sex-specific differences in the grey mouse lemur (*Microcebus murinus*). *Behavioral Ecology and Sociobiology* 50:493–502.

Ramakrishnan, U., and Coss, R. G. 2000. Recognition of heterospecific alarm vocalizations by bonnet macaques (*Macaca radiata*). *Journal of Comparative Psychology* 114:3–12.

Ramanankirahina, R., Joly, M., Scheumann, M., and Zimmermann, E. 2016. The role of acoustic signaling for spacing and group coordination in a nocturnal, pair-living primate, the western woolly lemur (*Avahi occidentalis*). *American Journal of Physical Anthropology* 159:466–477.

Ramos-Fernández, G., Mateos, J. L., Miramontes, O., Cocho, G., Larralde, H., and Ayala-Orozco, B. 2004. Lévy walk patterns in the foraging movements of spider monkeys (*Ateles geoffroyi*). *Behavioral Ecology and Sociobiology* 55:223–230.

Ramos-Fernández, G., Smith Agular, S. E., Schaffner, C. M., Vick, L. G., and Aureli, F. 2013. Site fidelity in space use by spider monkeys (*Ateles geoffroyi*) in the Yucatan Peninsula, Mexico. *PLoS ONE* 8:e62813. DOI: 10.1371/journal.pone.0062813

Ramsier, M. A., and Rauschecker, J. P. 2017. Primate audition: Reception, perception, and ecology. In *Springer Handbook of Auditory Research. Vol. 63. Primate Hearing and Communication* (Quam, R. M., Ramsier, M. A., Fay, R. R., and Popper, A. N., eds.). Springer, Cham, Switzerland, pp. 47–77.

Range, F., Forderer, T., Storrer-Meystre, Y., Benetton, C., and Fruteau, C. 2007. The structure of social relationships among sooty mangabeys in Taï. In *Monkeys of the Taï Forest: an African Primate Community* (McGraw, W. S., Zuberbühler, K., and Noë, R., eds.). Cambridge University Press, Cambridge, UK, pp. 109–130.

Rasa, O. A. E. 1987. The dwarf mongoose: a study of behavior and social structure in relation to ecology in a small, social carnivore. *Advances in the Study of Behavior* 17:121–162.

Reed, J. M., and Dobson, A. P. 1993. Conservation biology: Conspecific attraction and recruitment. *Trends in Ecology and Evolution* 8:253–256.

Reinhardt, K. D., Wirdateti, and Nekaris, K. A. I. 2016. Climate-mediated activity of the Javan slow loris, *Nycticebus javanicus*. *AIMS Environmental Science* 3:249–260.

Reyna-Hurtado, R., Teichroeb, J. A., Bonnell, T. R., Hernández-Sarabia, R. U., Vickers, S. M., Serio-Silva, J. C., Sicotte, P., and Chapman, C. A. 2018. Primates adjust movement strategies due to changing food availability. *Behavioral Ecology* 29:368–376.

Reynolds, A. M. 2018. Current status and future directions of Lévy walk research. *Biology Open* 7:bio030106. DOI: 10.1242/bio.030106

Reznick, D., and Bryant, M. 2007. Comparative long-term mark-recapture studies of guppies (*Peocilia reticulata*): Differences among high and low predation localities in growth and survival. *Annales Zoologici Fennici* 44:152–160.

Reznick, D., Butler, M. J., and Rodd, H. 2001. Life-history evolution in guppies. VII. The comparative ecology of high- and low-predation environments. *American Naturalist* 157:126–140.

Richard, A. 1970. A comparative study of the activity patterns and behavior of *Alouatta villosa* and *Ateles geoffroyi*. *Folia Primatologica* 12:241–263.

Riley, E. P. 2008. Ranging patterns and habitat use of Sulawesi Tonkean macaques (*Macaca tonkeana*) in a human-modified habitat. *American Journal of Primatology* 76:670–679.

Robbins, A. M., Stoinski, T., Fawcett, K., and Robbins, M. M. 2009a. Leave or conceive: natal dispersal and philopatry of female mountain gorillas in the Virunga volcano region. *Animal Behaviour* 77:831–838.

Robbins, A. M., Stoinski, T. S., Fawcett, K. A., and Robbins, M. M. 2009b. Does dispersal cause reproductive delays in female mountain gorillas? *Behaviour* 146:525–549.

Robbins, M. M., Robbins, A. M., Gerald-Steklis, N., and Steklis, H. D. 2007. Socieco-logical influences on the reproductive success of female mountain gorillas (*Gorilla beringei beringei*). *Behavioral Ecology and Sociobiology* 61:919–931.

Roberts, S.-J., and Cords, M. 2013. Group size but not dominance rank predicts the probability of conception in a frugivorous primate. *Behavioral Ecology and Sociobiology* 67:1995–2009.

Robinson, J. G. 1986. Seasonal variation in use of time and space by the wedge-capped capuchin monkey, *Cebus olivaceus*: Implications for foraging theory. *Smithsonian Contributions to Zoology* 431:1–60.

Rode, E. J., Nekaris, K. A. I., Markolf, M., Schliehe-Diecks, S., Seiler, M., Radespiel, U., and Schwitzer, C. 2013. Social organization of the northern giant mouse lemur *Mirza zaza* in Sahamalaza, northwestern Madagascar, inferred from nest group composition and genetic relatedness. *Contributions to Zoology* 82:71–83.

Rode-Margono, E. J., Nijman, V., Wirdateti, and Nekaris, K. A. I. 2014. Ethology of the critically endangered Javan slow loris *Nycticebus javanicus* É. Geoffroy Saint-Hilaire in West Java. *Asian Primates Journal* 4:27–41.

Rode-Margono, E. J., Schwitzer, C., and Nekaris, K. A.-I. 2016. Morphology, behaviour, ranging patterns and habitat use of the northern giant mouse lemur *Mirza zaza* in Sahamalaza, northwestern Madagascar. In *The Dwarf and Mouse Lemurs of Madagascar* (Lehman, S. M., Radespiel, U., and Zimmermann, E., eds.). Cambridge University Press, New York, pp. 235–254.

Rodman, P. S. 1973. Population composition and adaptive organization among orang-utans of the Kutai Reserve. In *Comparative Ecology and Behaviour of Primates: Proceedings of a Conference* (Michael, R. P., ed.). Academic Press, New York, pp. 171–209.

Rodman, P. S. 1979. Individual activity patterns and the solitary nature of orangutans. In *The Great Apes* (Hamburg, D. A., and McCown, E. R., ed.). Benjamin/Cummings Publishing Co., Menlo Park, CA, pp. 235–255.

Rodman, P. S. 1999. Whither primatology? The place of primates in contemporary anthropology. *Annual Review of Anthropology* 28:311–339.

Rodman, P. S., and McHenry, H. M. 1980. Bioenergetics and the origin of bipedalism. *American Journal of Physical Anthropology* 52:103–106.

Rogers, L. D., and Nekaris, K. A. I. 2011. Behaviour and habitat use of the Bengal slow loris *Nycticebus bengalensis* in the dry dipterocarp forests of Phnom Samkos Wildlife Sanctuary, Cambodia. *Cambodian Journal of Natural History* 2:104–113.

Roll, U., Dayan, T., and Kronfeld-Schor, N. 2006. On the role of phylogeny in determining activity patterns of rodents. *Evolutionary Ecology* 20:479–490.

Rose, L. M. 1994. Sex differences in diet and foraging behavior in white-faced capuchins (*Cebus capucinus*). *International Journal of Primatology* 15:95–114.

Rowe, N., and Myers, M. (eds.). 2016. *All the World's Primates*. Pogonias Press, Charlestown, RI.

Ruivo, L. V. P., Stone, A. I., and Fienup, M. 2017. Reproductive status affects the feeding ecology and social association patterns of female squirrel monkeys (*Saimiri collinsi*) in an Amazonian rainforest. *American Journal of Primatology* 79:e22657. DOI: 10.1002/ajp.22657

Rylands, A. B. 1989. Sympatric Brazilian callitrichids: the black tufted-ear marmoset, *Callithrix kuhli*, and the golden-headed lion tamarin, *Leontopithecus chrysomelas*. *Journal of Human Evolution* 18:679–695.

Sakamaki, T., Behncke, I., Laporte, M., Mulavwa, M., Ryu, H., Takemoto, H., Tokuyama, N., Yamamoto, S., and Furuichi, T. 2015. Intergroup transfer of females and social relationships between immigrants and residents in bonobo (*Pan paniscus*) societies. In *Dispersing Primate Females: Life History and Social Strategies in Male-Philopatric Species* (Furuichi, T., Yamagiwa, J., and Aureli, F., eds.). Springer Japan, Tokyo, pp. 127–164.

Sakamaki, T., Ryu, H., Toda, K., Tokuyama, N., and Furuichi, T. 2018. Increased frequency of intergroup encounters in wild bonobos (*Pan paniscus*) around the yearly peak in fruit abundance at Wamba. *International Journal of Primatology* 39:685–704.

Sato, H. 2012. Diurnal resting in brown lemurs in a dry deciduous forest, northwestern Madagascar: Implications for seasonal thermoregulation. *Primates* 53:255–263.

Sauther, M. L., Sussman, R. W., and Gould L. 1999. The socioecology of the ringtailed lemur: Thirty-five years of research. *Evolutionary Anthropology* 8:120–132.

Sayers, K., and Norconk, M. A. 2008. Himalayan *Semnopithecus entellus* at Langtang National Park, Nepal: Diet, activity patterns, and resources. *International Journal of Primatology* 29:509–530.

Scarry, C. J., Salmi, R., Lodwick, J., and Doran-Sheehy, D. M. 2023. Long-term home range stability provides foraging benefits in western gorillas. *American Journal of Biological Anthropology* 181:296–311.

Schmelting, B., Zimmermann, E., Berke, O., Bruford, M. W., and Radespiel, U. 2007. Experience-dependent recapture rates and reproductive success in male grey mouse lemurs (*Microcebus murinus*). *American Journal of Physical Anthropology* 133:743–752.

Schreier, A. L., Bolt, L. M., Russell, D. G., Readyhough, T. S., Jacobson, Z. S., Merrigan-Johnson, C., and Coggeshall, E. M. C. 2021. Mantled howler monkeys (*Alouatta palliata*) in a Costa Rican forest fragment do not modify activity budgets or spatial cohesion in response to anthropogenic edges. *Folia Primatologica* 92:49–57.

Schreier, A. L., and Grove, M. 2010. Ranging patterns of hamadryas baboons: Random walk analyses. *Animal Behaviour* 80:75–87.

Schülke, O. 2001. Social anti-predator behaviour in a nocturnal lemur. *Folia Primatologica* 72:332–334.

Schülke, O. 2003. To breed or not to breed—food competition and other factors involved in female breeding decisions in the pair-living nocturnal fork-marked lemur (*Phaner furcifer*). *Behavioral Ecology and Sociobiology* 55:11–21.

Schülke, O. 2005. Evolution of pair-living in *Phaner furcifer*. *International Journal of Primatology* 26:903–919.

Schülke, O., and Kappeler, P. M. 2003. So near and yet so far: Territorial pairs but low cohesion between pair partners in a nocturnal lemur, *Phaner furcifer*. *Animal Behaviour* 65:331–342.

Seiler, M., Holderied, M., and Schwitzer, C. 2014. Habitat selection and use in the critically endangered Sahamalaza sportive lemur *Lepilemur sahamalazensis* in altered habitat. *Endangered Species Research* 24:273–286.

Sekulic, R. 1982a. Behavior and ranging patterns of a solitary female red howler (*Alouatta seniculus*). *Folia Primatologica* 38:217–232.

Sekulic, R. 1982b. Daily and seasonal patterns of roaring and spacing in four red howler *Alouatta seniculus* troops. *Folia Primatologica* 39:22–48.

Seyfarth, R. M., and Cheney, D. L. 1980. The ontogeny of vervet monkey alarm calling behavior: a preliminary report. *Zietschrift für Tierpsychologie* 54:37–56.

Seyfarth, R. M., Cheney, D. L., and Marler, P. 1980a. Monkey responses to three different alarm calls: Evidence of predator classification and semantic communication. *Science* 210:801–803.

Seyfarth, R. M., Cheney, D. L., and Marler, P. 1980b. Vervet monkey alarm calls: Semantic communication in a free-ranging primate. *Animal Behaviour* 28:1070–1094.

Shaffer, C. A. 2013. Activity patterns, intergroup encounters, and male affiliation in free-ranging bearded sakis (*Chiropotes sagulatus*). *International Journal of Primatology* 34:1190–1208.

Shaffer, C. A. 2014. Spatial foraging in free ranging bearded sakis: Traveling salesmen or Lévy walkers? *American Journal of Primatology* 76:472–484.

Shanee, S., Allgas, N., and Shanee, N. 2013. Preliminary observations on the behavior and ecology of the Peruvian night monkey (*Aotus miconax*: Primates) in a remnant cloud forest patch, north eastern Peru. *Tropical Conservation Science* 6:1380148.

Shanee, S., and Shanee, N. 2011. Activity budget and behavioural patterns of free-ranging yellow-tailed woolly monkeys *Oreonax flavicauda* (Mammalia: Primates), at La Esperanza, northeastern Peru. *Contributions to Zoology* 80:269–277.

Shattuck, M., and Williams, S. A. 2010. Arboreality has allowed for the evolution of increased longevity in mammals. *Proceedings of the National Academy of Sciences USA* 107:4635–4639.

Shelley, E. L., and Blumstein, D. T. 2005. The evolution of vocal alarm communication in rodents. *Behavioral Ecology* 16:169–177.

Shimooka, Y. 2005. Sexual differences in ranging of *Ateles belzebuth* at La Macarena, Colombia. *International Journal of Primatology* 26:385–406.

Shultz, S., Opie, C., and Atkinson, Q. D. 2011. Stepwise evolution of stable sociality in primates. *Nature* 479:219–224.

Sicotte, P., and MacIntosh, A. J. 2004. Inter-group encounters and male incursions in *Colobus vellerosus* in centra Ghana. *Behaviour* 141:533–553.

Silk, J. B. 2012. The adaptive value of sociality. In *The Evolution of Primate Societies* (Mitani, J. C., Call, J., Kappeler, P. M., Palombit, R. A., and Silk, J. B., eds.). University of Chicago Press, Chicago, IL, pp. 552–551.

Simmen, B., Bayart, F., Rasamimanana, H., Zahariev, A., Blanc, S., and Pasquet, P. 2010. Total energy expenditure and body composition in two free-living sympatric lemurs. *PLoS One* 5:e9860. DOI: 10.1371/journal.pone.000986

Sims, D. W., Reynolds, A. M., Humphries, N. E., Southall, E. J., Wearmouth, V. J., Metcalfe, B., and Twitchett, R. J. 2014. Hierarchical random walks in trace fossils and the origin of optimal search behavior. *Proceedings of the National Academy of Sciences USA* 111:11073–11078.

Sinclair, A. R. E., Mduma, S., and Brashares, J. S. 2003. Patterns of predation in a diverse predator-prey system. *Nature* 425:288–290.

Singleton, I., and van Schaik, C. P. 2001. Orangutan home range size and its determinants in a Sumatran swamp forest. *International Journal of Primatology* 22:877–911.

Singleton, I., and van Schaik, C. P. 2002. The social organization of a population of Sumatran orang-utans. *Folia Primatologica* 73:1–20.

Snaith, T. V., and Chapman, C. A. 2008. Red colobus monkeys display alternative behavioral responses to the costs of scramble competition. *Behavioral Ecology* 19:1289–1296.

Snaith, T. V., Chapman, C. A., Rothman, J. M., and Wasserman, M. D. 2008. Bigger groups have fewer parasites and similar cortisol levels: a multi-group analysis in red colobus monkeys. *American Journal of Primatology* 70:1072–1080.

Soares, S. C., and Esteves, F. 2013. A glimpse of fear: Fast detection of threatening targets in visual search with brief stimulus durations. *PsyCh Journal* 2:11–16. DOI: 10.1002/pchj.18

Soares, S. C., Lindström, B., Esteves, F., and Öhman, A. 2014. The hidden snake in the grass: Superior detection of snakes in challenging attentional conditions. *PLoS ONE* 9:e114724. DOI: 10.1371/journal.pone.0114724

Soares, S. C., Maior, R. S., Isbell, L. A., Tomaz, C., and Nishijo, H. 2017. Fast detector/first responder: interactions between the superior colliculus-pulvinar pathway and stimuli relevant to primates. *Frontiers in Neuroscience: Systems Biology* 11:67. DOI: 10.3389/fnins.2017.00067

Sockol, M. D., Raichlen, D. A., and Pontzer, H. 2007. Chimpanzee locomotor energetics and the origin of human bipedalism. *Proceedings of the National Academy of Sciences USA* 104:12265–12269.

Souza-Alves, J. P., Chagas, R. R. D., Santana, M. M., Boyle, S. A., and Bezerra, B. M. 2021a. Food availability, plant diversity, and vegetation structure drive behavioral and ecological variation in endangered Coimbra-Filho's titi monkeys. *American Journal of Primatology* 83:e23237. DOI: 10.1002/ajp.23237

Souza-Alves, J. P., Chagas Alves, R. R. D., Hilário, R. R., Barnett, A. A., and Bezerra, B. M. 2021b. Species-specific resource availability as potential correlates of foraging strategy in Atlantic Forest edge-living common marmosets. *Ethology Ecology and Evolution* 34:449–470.

Spence-Aizenberg, A., Di Fiore, A., and Fernandez-Duque, E. 2016. Social monogamy, male-female relationships, and biparental care in wild titi monkeys (*Callicebus discolor*). *Primates* 57:103–112.

Städele, V., Van Doren, V., Pines, M., Swedell, L., and Vigilant, L. 2015. Fine-scale genetic assessment of sex-specific dispersal patterns in a multilevel primate society. *Journal of Human Evolution* 78:103–113.

Stanford, C. B. 1995. The influence of chimpanzee predation on group size and anti-predator behaviour in red colobus monkeys. *Animal Behaviour* 49:577–587.

Stanford, C. B. 1998. *Chimpanzee and Red Colobus: The Ecology of Predator and Prey.* Harvard University Press, Cambridge, MA.

Stanford, C. B., Wallis, J., Matama, H., and Goodall, J. 1994. Patterns of predation by chimpanzees on red colobus monkeys in Gombe National Park, 1982–1991. *American Journal of Physical Anthropology* 94:213–228.

Stankowich, T., and Coss, R. G. 2008. Alarm walking in Columbian black-tailed deer: Its characterization and possible antipredator signaling functions. *Journal of Mammalogy* 89:636–645.

Stearns, S. C. 1992. *The Evolution of Life Histories.* Oxford University Press, New York.

Steenbeek, R., and van Schaik, C. P. 2001. Competition and group size in Thomas's langurs (*Presbytis thomasi*): The folivore paradox revisited. *Behavioral Ecology and Sociobiology* 49:100–110.

Sterck, E. H. M. 1997. Determinants of female dispersal in Thomas langurs. *American Journal of Primatology* 42:179–198.

Sterck, E. H. M., and van Hooff, J. A. R. A. M. 2000. The number of males in langur groups: Monopolizability of females or demographic processes? In *Primate Males: Causes and Consequences of Variation in Group Composition* (Kappeler, P. M., ed.). Cambridge University Press, Cambridge, UK, pp. 120–129.

Sterck, E. H. M., Watts, D. P., and van Schaik, C. P. 1997. The evolution of female social relationships in nonhuman primates. *Behavioral Ecology and Sociobiology* 41:291–309.

Sterck, E. H. M., Willems, E. P., van Hooff, J. A. R. A. M., and Wich, S. A. 2005. Female dispersal, inbreeding avoidance and mate choice in Thomas langurs (*Presbytis thomasi*). *Behaviour* 142:845–868.

Stevenson, P. R., Quiñones, M. J., and Ahumada, J. A. 1994. Ecological strategies of woolly monkeys (*Lagothrix lagotricha*) at Tinigua National Park, Colombia. *American Journal of Primatology* 32:13–140.

Stevenson, P. R., Quinones, M. J., and Ahumada, J. A. 2000. Influence of fruit availability on ecological overlap among four Neotropical primates at Tinigua National Park, Colombia. *Biotropica* 32:533–544.

Stevenson, P. R., Zárate, D. A., Ramírez, M. A., and Henao-Díaz, F. 2015. Social interactions and proximal spacing in woolly monkeys: Lonely females looking for male friends. In *Dispersing Primate Females: Life History and Social Strategies in Male-Philopatric Species* (Furuichi, T., Yamagiwa, J., and Aureli, F., eds.). Springer Japan, Tokyo, pp. 45–71.

Strauss, W. M., Hetem, R. S., Mitchell, D., Maloney, S. K., O'Brien, H. D., Meyer. L. C. R., and Fuller, A. 2017. Body water conservation through selective brain cooling by the carotid rete: A physiological feature for surviving climate change? *Conservation Physiology* 5:cow078. DOI: 10.1093/conphys/cow078

Strier, K. B. 1987. Activity budgets of woolly spider monkeys, or muriquis (*Brachyteles arachnoides*). *American Journal of Primatology* 13:385–395.

Strier, K. B. 1997. Mate preferences of wild muriqui monkeys (*Brachyteles arachnoides*): reproductive and social correlates. *Folia Primatologica* 68:120–133.

Strier, K. B., and Ives, A. R. 2012. Unexpected demography in the recovery of an endangered primate population. *PLoS ONE* 7:e44407. DOI: 10.1371/journal.pone.004407

Strier, K. B., Possamai, C. B., and Mendes, S. L. 2015. Dispersal patterns of female northern muriquis: Implications for social dynamics, life history, and conservation. In *Dispersing Primate Females: Life History and Social Strategies in Male-Philopatric Species* (Furuichi, T., Yamagiwa, J., and Aureli, F., eds.). Springer Japan, Tokyo, pp. 3–22.

Strier, K. B., and Ziegler, T. E. 2000. Lack of pubertal influences on female dispersal in muriqui monkeys, *Brachyteles arachnoides*. *Animal Behaviour* 59:849–860.

Struhsaker, T. T. 1967. Ecology of vervet monkeys (*Cercopithecus aethiops*) in the Masai-Amboseli Game Reserve, Kenya. *Ecology* 48:891–904.

Struhsaker, T. T. 1969. Correlates of ecology and social organization among African cercopithecines. *Folia Primatologica* 11:80–118.

Struhsaker, T. T. 1975. *The Red Colobus Monkey.* University of Chicago Press, Chicago, IL.

Struhsaker, T. T. 2000. Variation in adult sex ratios of red colobus monkey social groups: Implications for interspecific comparisons. In *Primate Males: Causes and Consequences of Variation in Group Composition* (Kappeler, P. M., ed.). Cambridge University Press, Cambridge, UK, pp. 108–119.

Struhsaker, T. T. 2010. *The Red Colobus Monkeys: Variation in Demography, Behavior, and Ecology of Endangered Species.* Oxford University Press, New York.

Struhsaker, T. T., and Leland, L. 1979. Socioecology of five sympatric monkey species in the Kibale Forest, Uganda. *Advances in the Study of Behavior* 9:159–228.

Struhsaker, T. T., and Oates, J. F. 1975. Comparison of the behavior and ecology of red colobus and black-and-white colobus monkeys in Uganda: A summary. In *Socioecology and Psychology of Primates* (Tuttle, R. H., ed.). De Gruyter Mouton, The Hague, pp. 103–124.

Suarez, S. A. 2006. Diet and travel costs for spider monkeys in a nonseasonal, hyperdiverse environment. *International Journal of Primatology* 27:411–436.

Sueur, C. 2011. A non-Lévy random walk in chacma baboons: What does it mean? *PLoS ONE* 6:e16131. DOI: 10.1371/journal.pone.0016131

Sugardjito, J., te Boekhorst, I. J. A., and van Hooff, J. A. R. A. M. 1987. Ecological constraints on the grouping of wild orang-utans (*Pongo pygmaeus*) in the Gunung Leuser National Park, Sumatra, Indonesia. *International Journal of Primatology* 8:17–41.

Sugiura, H., Shimooka, Y., and Tsuji, Y. 2013. Japanese macaques depend not only on neighbors but also on more distant members for group cohesion. *Ethology* 120:21–31.

Sussman, R. W. 1992. Primate origins and the evolution of angiosperms. *American Journal of Primatology* 23:209–223.

Suzuki, M., and Sugiura, H. 2011. Effects of proximity and activity on visual and auditory monitoring in wild Japanese macaques. *American Journal of Primatology* 73:623–631.

Swapna, N., Radhakrishna, S., Gupta, A. K., and Kumar, A. 2010. Exudativory in the Bengal slow loris (*Nycticebus bengalensis*) in Trishna Wildlife Sanctuary, Tripura, Northeast India. *American Journal of Primatology* 72:113–121.

Swedell, L. 2002. Ranging behavior, group size and behavioral flexibility in Ethiopian hamadryas baboon (*Papio hamadryas hamadryas*). *Folia Primatologica* 75:95–103.

Swedell, L., Saunders, J., Schreier, A., Davis, B., Tesfaye, T., and Pines, M. 2011. Female "dispersal" in hamadryas baboons: Transfer among social units in a multilevel society. *American Journal of Physical Anthropology* 145:360–370.

Symington, M. M. 1988. Demography, ranging patterns, and activity budgets of black spider monkeys (*Ateles paniscus chamek*) in the Manu National Park, Peru. *American Journal of Primatology* 15:45–67.

Tabacow, F. P., Mendes, S. L., and Strier, K. B. 2009. Spread of a terrestrial tradition in an arboreal primate. *American Anthropologist* 111:238–249.

Takahashi, M. Q., Rothman, J. M., Raubenheimer, D., and Cords, M. 2019. Dietary generalists and nutritional specialists: Feeding strategies of adult female blue monkeys (*Cercopithecus mitis*) in the Kakamega Forest, Kenya. *American Journal of Primatology* 81:e23016. DOI: 10.1002/ajp.23016

Takasaki, H. 1981. Troop size, habitat quality, and home range area in Japanese macaques. *Behavioral Ecology and Sociobiology* 9:277–281.

Talebi, M. G., and Lee, P. C. 2010. Activity patterns of *Brachyteles arachnoides* in the largest remaining fragment of Brazilian Atlantic forest. *International Journal of Primatology* 31:571–583.

Taylor, C. R., and Rowntree, V. J. 1973. Running on two or four legs: which consumes more energy? *Science* 179:186–187.

Tecot, S. R., Singletary, B., and Eadie, E. 2016. Why "monogamy" isn't good enough. *American Journal of Primatology* 78:340–354.

Teelen, S. 2008. Influence of chimpanzee predation on the red colobus population at Ngogo, Kibale National Park, Uganda. *Primates* 49:41–49.

Teichroeb, J. A., Saj, T. L., Paterson, J. D., and Sicotte, P. 2003. Effect of group size on activity budgets of *Colobus vellerosus* in Ghana. *International Journal of Primatology* 24:743–758.

Teichroeb, J. A., and Sicotte, P. 2009. Test of the ecological constraints model in ursine colobus monkeys (*Colobus vellerosus*) in Ghana. *American Journal of Primatology* 71:49–59.

Teichroeb, J. A., Wikberg, E. C, and Sicotte, P. 2009. Female dispersal patterns in six groups of ursine colobus (*Colobus vellerosus*): Infanticide avoidance is important. *Behaviour* 146:551–582.

Terborgh, J. 1983. *Five New World Primates: A Study in Comparative Ecology*. Princeton University Press, Princeton, NJ.

Terborgh, J. and Janson, C. H. 1986. The socioecology of primate groups. *Annual Review of Ecology and Systematics* 17:111–135.

Thierry, B. 2008. Primate socioecology, the lost dream of ecological determinism. *Evolutionary Anthropology* 17:93–96.

Thierry, B. 2013. Identifying constraints in the evolution of primate societies. *Philosophical Transactions of the Royal Society B* 368:20120342. DOI: 10/1098/rstb .2012.0342

Thompson, C. L. 2016. To pair or not to pair: Sources of social variability with white-faced saki monkeys (*Pithecia pithecia*) as a case study. *American Journal of Primatology* 78:561–572.

Thompson, C. L., and Norconk, M. A. 2011. Within-group social bonds in white-faced saki monkeys (*Pithecia pithecia*) display male-female pair preference. *American Journal of Primatology* 73:1051–1061.

Thompson, C. L., Williams, S. H., Glander, K. E., and Vinyard, C. J. 2016. Measuring microhabitat temperature in arboreal primates: A comparison of on-animal and stationary approaches. *International Journal of Primatology* 37:495–517.

Thorén, S., Quietzsch, F., Schwochow, D., Sehen, L., Meusel, C., Meares, K., and Radespiel, U. 2011. Seasonal changes in feeding ecology and activity patterns of two sympatric mouse lemur species, the gray mouse lemur (*Microcebus murinus*) and the golden-brown mouse lemur (*M. ravelobensis*), in northwestern Madagascar. *International Journal of Primatology* 32:566–586.

Thorington, R. W., Jr., Koprowski, J. L., Steele, M. A., and Watton, J. F. 2012. *Squirrels of the World*. The Johns Hopkins University Press, Baltimore, MD.

Thorpe, S. K. S., Crompton, R. H., and Alexander, R. McN. 2007. Orangutans use compliant branches to lower the energetic cost of locomotion. *Biology Letters* 3:253–256.

Thorpe, S. K. S., Holder, R., and Crompton, R. H. 2009. Orangutans employ unique strategies to control branch flexibility. *Proceedings of the National Academy of Sciences USA* 106:12646–12651.

Toda, K., Mouri, K., Ryu, H., Sakamaki, T., Tokuyama, N., Yokoyama, T., Shibata, S., Poiret, M.-L., Shimizu, K., Hashimoto, C., and Furuichi, T. 2022. Do female bonobos (*Pan paniscus*) disperse at the onset of puberty? *Hormones and Behavior* 142:105159. DOI: 10.1016/j.yhbeh.2022.105159

Tracy, C. R., and Christian, K. A. 1986. Ecological relations among space, time, and thermal niche axes. *Ecology* 67:609–615.

Treves, A. 1999. Has predation shaped the social systems of arboreal primates? *International Journal of Primatology* 20:35–67.

Treves, A. 2000. Theory and method in studies of vigilance and aggregation. *Animal Behaviour* 60:711–722.

Treves, A. 2001. Reproductive consequences of variation in the composition of howler monkey (*Alouatta* spp.) groups. *Behavioral Ecology and Sociobiology* 50:61–71.

Trivers, R. L. 1972. Parental investment and sexual selection. In *Sexual Selection and the Descent of Man, 1871–1971* (Campbell, B., ed.). Aldine, Chicago, IL, pp. 136–179.

Turner, M. I. M., and Watson, R. M. 1965. An introductory study on the ecology of hyrax (*Dendrohyrax brucei* and *Procavia johnstoni*) in the Serengeti National Park. *East African Wildlife Journal* 3:49–60

Ulibarri, L. R., and Gartland, K. 2021. Group composition and social structure of red-shanked douc (*Pygathrix nemaeus*) at Son Tra Nature Reserve, Vietnam. *Folia Primatologica* 92:191–202.

Valero, A., and Byrne, R. W. 2007. Spider monkey ranging patterns in Mexican subtropical forest: Do travel routes reflect planning? *Animal Cognition* 10:305–315.

Vanaraj, G., and Pragasan, L. A. 2021. Activity and dietary budgets of tufted grey langurs (*Semnopithecus priam priam*). *Ethology Ecology and Evolution* 33:477–495.

Van Belle, S., and Estrada, A. 2008. Group size and composition influence male and female reproductive success in black howler monkeys (*Alouatta pigra*). *American Journal of Primatology* 70:613–619.

Van Belle, S., and Estrada, A. 2020. The influence of loud calls on intergroup spacing mechanism in black howler monkeys (*Alouatta pigra*). *International Journal of Primatology* 41:265–286.

Van Belle, S., Fernandez-Duque, E., and Di Fiore, A. 2016. Demography and life history of wild red titi monkeys (*Callicebus discolor*) and equatorial sakis (*Pithecia aequatorialis*) in Amazonian Ecuador: A 12-year study. *American Journal of Primatology* 78:204–215.

Van Belle, S., Porter, A., Fernandez-Duque, E., and Di Fiore, A. 2018. Ranging behavior and potential for territoriality in equatorial sakis (*Pithecia aequatorialis*) in Amazonian Ecuador. *American Journal of Physical Anthropology* 167:701–712.

Van Belle, S., Porter, A., Fernandez-Duque, E., and Di Fiore, A. 2020. Ranging behavior and the potential for territoriality in pair-living titi monkeys (*Plecturocebus discolor*). *American Journal of Primatology* 83:e23225. DOI: 10.1002/ajp.23225

Vandercone, R., Premachandra, K., Wijethunga, G. P, Dinadh, C., Ranawana, K., and Bahar, S. 2013. Random walk analysis of ranging patterns of sympatric langurs in a complex resource landscape. *American Journal of Primatology* 75:1209–1219.

van Noordwijk, M. A., Arora, N., Willems, E. P., Dunkel, L. P., Amda, R. N., Mardianah, N., Ackermann, C., Krützen, M., and van Schaik, C. P. 2012. Female philopatry and its social benefits. *Behavioral Ecology and Sociobiology* 66:823–834.

van Schaik, C. P. 1983. Why are diurnal primates living in groups? *Behaviour* 87: 120–144.

van Schaik, C. P. 1989. The ecology of social relationships amongst female primates. In *Comparative Socioecology: The Behavioural Ecology of Humans and Other Mammals* (Standen, V., and Foley, R. A., eds.). Blackwell Scientific Publications, London, pp. 195–218.

van Schaik, C. P. 1996. Social evolution in primates: The role of ecological factors. *Proceedings of the British Academy* 88:9–31.

van Schaik, C. P. 1999. The socioecology of fission-fusion sociality in orangutans. *Primates* 40:69–86.

van Schaik, C. P., Assink, P., and Salafsky, N. 1992. Territorial behavior in Southeast Asian langurs: Resource defense or mate defense? *American Journal of Primatology* 26:233–242.

van Schaik, C. P., and Dunbar, R. I. M. 1990. The evolution of monogamy in large primates: A new hypothesis and some crucial tests. *Behaviour* 115:30–62.

van Schaik, C. P., and Mitrasetia, T. 1990. Changes in the behaviour of wild long-tailed macaques (*Macaca fascicularis*) after encounters with a model python. *Folia Primatologica* 55:104–108.

van Schaik, C. P., and van Hooff, J. A. R. A. M. 1983. On the ultimate causes of primate social systems. *Behaviour* 85:91–117.

van Schaik, C. P., van Noordwijk, M. A., de Boer, R. J., and den Tonkelaar, I. 1983a. The effect of group size on time budgets and social behaviour in wild long-tailed macaques (*Macaca fascicularis*). *Behavioral Ecology and Sociobiology* 13:173–181.

van Schaik, C. P., van Noordwijk, M. A., Warsono, B., and Sutriono, E. 1983b. Party size and early detection of predators in Sumatran forest primates. *Primates* 24:211–221.

van Strien, J. W., and Isbell, L. A. 2017. Snake scales, partial exposure, and the Snake Detection Theory: A human ERP study. *Scientific Reports* 7:46331. DOI:10.1038/srep46331

Vasey, N. 2005. Activity budgets and activity rhythms in red ruffed lemurs (*Varecia rubra*) on the Masoala Peninsula, Madagascar: Seasonality and reproductive energetics. *American Journal of Primatology* 66:23–44.

Vick, L. G., and Pereira, M. E. 1989. Episodic targeting aggression and the histories of *Lemur* social groups. *Behavioral Ecology and Sociobiology* 25:3–12.

Viswanathan, G. M., Buldyrev, S. V., Havlin, S., da Luz, M. G., Raposo, E. P., and Stanley, H. E. 1999. Optimizing the success of random searches. *Nature* 401:911–914.

Vogel, E. R., and Dominy, N. J. 2011. Measuring ecological variables for primate field studies. In *Primates in Perspective, 2nd edition* (Campbell, C. J., Fuentes, A., MacKinnon, K. C., Bearder, S. K., and Stumpf, R. M., eds.). Oxford University Press, New York, pp. 367–377.

Vogel, E. R., and Janson, C. H. 2011. Quantifying primate food distribution and abundance for socioecological studies: An objective consumer-centered method. *International Journal of Primatology* 32:737–754.

Walker, K. K., Walker, C. S., Goodall, J., and Pusey, A. E. 2018. Maturation is prolonged and variable in female chimpanzees. *Journal of Human Evolution* 114:131–140.

Wallace, R. B. 2001. Diurnal activity budgets of black spider monkeys, *Ateles chamek*, in a southern Amazonian tropical forest. *Neotropical Primates* 9:101–107.

Wallace, R. B. 2008. Towing the party line: territoriality, risky boundaries and male group size in spider monkey fission-fusion societies. *American Journal of Primatology* 70:271–281.

Wallen, M. M., Patterson, E. M., Krzyszczyk, E., and Mann, J. 2016. The ecological costs to females in a system with allied sexual coercion. *Animal Behaviour* 115:227–236.

Walther, F. R. 1969. Flight behaviour and avoidance of predators in Thomson's gazelle (*Gazella thomsoni* Guenther 1884). *Behaviour* 34:184–221.

Ward, P., and Zahavi, A. 1973. The importance of certain assemblages of birds as "information-centres" for food-finding. *Ibis* 115:517–534.

Wartmann, F. M., Juárez, C. P., and Fernandez-Duque, E. 2014. Size, site fidelity, and overlap of home ranges and core areas in the socially monogamous owl monkey (*Aotus azarae*) of northern Argentina. *International Journal of Primatology* 35:919–939.

Wartmann, F. M., Purves, R. S., and van Schaik, C. P. 2010. Modeling ranging behaviour of female orang-utans: a case study in Tuanan, Central Kalimantan, Indonesia. *Primates* 51:119–130.

Waser, P. M. 1977. Feeding ranging and group size in the mangabey *Cercocebus albigena*. In *Primate Ecology: Studies of Feeding and Ranging Behaviour in Lemurs, Monkeys, and Apes* (Clutton-Brock, T. H., ed.). Academic Press, New York, pp. 183–222.

Waser, P. M., and Jones, W. T. 1983. Natal philopatry among solitary mammals. *Quarterly Review of Biology* 58:355–390.

Watts, D. P. 1988. Environmental influences on mountain gorilla time budgets. *American Journal of Primatology* 15:195–211.

Watts, D. P. 1989. Infanticide in mountain gorillas—new cases and a reconsideration of the evidence. *Ethology* 81:1–18.

Watts, D. P. 1994. Agonistic relationships between female mountain gorillas (*Gorilla gorilla beringei*). *Behavioral Ecology and Sociobiology* 34:347–358.

Watts, D. P. 1998. Long-term habitat use by mountain gorillas (*Gorilla gorilla beringei*). 1. Consistency, variation, and home range size and stability. *International Journal of Primatology* 19:651–680.

Watts, D. P., and Amsler, S. J. 2013. Chimpanzee-red colobus encounter rates show a red colobus population decline associated with predation by chimpanzees at Ngogo. *American Journal of Primatology* 75:927–937.

Watts, D. P., and Mitani, J. C. 2002. Hunting behavior of chimpanzees at Ngogo, Kibale National Park, Uganda. *International Journal of Primatology* 23:1–28.

Weidt, A., Hagenah, N., Randrianambinina, B., Radespiel, U., and Zimmermann, E. 2004. Social organization of the golden brown mouse lemur (*Microcebus ravelobensis*). *American Journal of Physical Anthropology* 123:40–51.

Welman, S., Tuen, A. A., and Lovegrove, B. G. 2017. Searching for the haplorrhine heterotherm: Field and laboratory data of free-ranging tarsiers. *Frontiers in Physiology* 8:745. DOI: 10.3389/fphys.2017.00745

Wessling, E. G., Kühl, H. S., Mundry, R., Deschner, T., and Pruetz, J. D. 2018. The costs of living at the edge: Seasonal stress in wild savanna-dwelling chimpanzees. *Journal of Human Evolution* 121:1–11.

White, F. J. 1992. Activity budgets, feeding behavior, and habitat use of pygmy chimpanzees at Lomako, Zaire. *American Journal of Primatology* 26:215–223.

Whiten, A., Byrne, R. W., and Henzi, S. P. 1987. The behavioral ecology of mountain baboons. *International Journal of Primatology* 8:367–388.

Wieczkowski, J. 2005. Examination of increased annual range of a Tana Mangabey (*Cercocebus galeritus*) group. *American Journal of Physical Anthropology* 128:381–388.

Wiens, F., and Zitzmann, A. 2003. Social structure of the solitary slow loris *Nycticebus coucang*. *Journal of Zoology, London* 261:35–46.

Willems, E. P., and Hill, R. A. 2009. Predator-specific landscapes of fear and resource distribution: Effects on spatial range use. *Ecology* 90:546–555.

Williams, B. A., Kay, R. F., and Kirk, E. C. 2010. New perspectives on anthropoid origins. *Proceedings of the National Academy of Sciences USA* 107:4797–4804.

Williams, G. C. 1966. *Adaptation and Natural Selection.* Princeton University Press, Princeton, NJ.

Williams, J. M., Oehlert, G. W., Carlis, J. V., and Pusey, A, E. 2004. Why do male chimpanzees defend a group range? *Animal Behaviour* 68:523–532.

Williams, J. M., Pusey, A. E, Carlis, J. V., Farm, B. P., and Goodall, J. 2002. Female competition and male territorial behaviour influence female chimpanzees' ranging patterns. *Animal Behaviour* 63:347–360.

Wilson, E. O. 1975. *Sociobiology: The New Synthesis.* Belknap Press, Cambridge, MA.

Wilson, M. L., Wallauer, W. R., and Pusey, A. E. 2004. New cases of intergroup violence among chimpanzees in Gombe National Park, Tanzania. *International Journal of Primatology* 25:523–549.

Wong, S. N. P., and Sicotte, P. 2007. Activity budget and ranging patterns of *Colobus vellerosus* in forest fragments in Central Ghana. *Folia Primatologica* 78:245–254.

Woodland, D. J., Jaafar, Z., and Knight, M.-L. 1980. The "pursuit deterrent" function of alarm signals. *American Naturalist* 115:748–753.

Woodroffe, R. 2011. Demography of a recovering wild dog (*Lycaon pictus*) population. *Journal of Mammalogy* 92:305–315.

Woodroffe, R., Davies-Mostert, H., Ginsberg, J., Graf, H., Leigh, K., McCreery, K., Mills, G., Pole, A., Rasmussen, G., Robbins, R., Somers, S., and Szykman, M. 2007. Rates and causes of mortality in endangered wild dogs *Lycaon pictus*: Lessons for management and monitoring. *Oryx* 41:215–223.

Wrangham, R. W. 1977. Feeding behaviour of chimpanzees in Gombe National Park, Tanzania. In *Primate Ecology: Studies of Feeding and Ranging Behaviour in Lemurs, Monkeys, and Apes* (Clutton-Brock, T. H., ed.). Academic Press, New York, pp. 504–538.

Wrangham, R. W. 1979. Sex differences in chimpanzee dispersion. In *The Great Apes* (Hamburg, D. A., and McCown, E. R., eds.). Benjamin/Cummings, San Francisco, CA, pp. 481–489.

Wrangham, R. W. 1980. An ecological model of female-bonded primate groups. *Behaviour* 75:262–300.

Wrangham, R. W. 1987. Evolution of social structure. In *Primate Societies* (Smuts, B. B., Cheney, D. L., Seyfarth, R. M., Wrangham, R. W., and Struhsaker, T. T., eds.). University of Chicago Press, Chicago, IL, pp. 282–298.

Wrangham, R. W. 2000. Why are male chimpanzees more gregarious than mothers? A scramble competition hypothesis. In *Primate Males: Causes and Consequences of Variation in Group Composition* (Kappeler, P. M., ed.). Cambridge University Press, Cambridge, UK, pp. 248–258.

Wright, P. C. 1995. Demography and life history of free-ranging *Propithecus diadema edwardsi* in Ranomafana National Park, Madagascar. *International Journal of Primatology* 16:835–854.

Xiang, Z., Huo, S., and Xiao, W. 2010. Activity budget of *Rhinopithecus bieti* at Tibet: Effects of day length, temperature and food availability. *Current Zoology* 56:650–659.

Yepez, P., de la Torre, S., and Snowdon, C. T. 2005. Interpopulation differences in exudate feeding of pygmy marmosets in Ecuadorian Amazonia. *American Journal of Primatology* 66:145–158.

Youlatos, D. 2009. Locomotion, postures, and habitat use by pygmy marmosets (*Cebuella pygmaea*). In *The Smallest Anthropoids: The Marmoset/Callimico Radiation* (Ford, S. M., Porter, L. M., and Davis, L. C., eds.). Springer, New York, pp. 279–300.

Young, C., McFarland, R., Ganswindt, A., Young, M. M. I., Barrett, L., and Henzi, S. P. 2019. Male residency and dispersal triggers in a seasonal breeder with influential females. *Animal Behaviour* 154:29–37.

Young, T. P. 1981. A general model of comparative fecundity for semelparous and iteroparous life histories. *American Naturalist* 118:27–36.

Young, T. P. 1992. The evolution of semelparity in Mount Kenya lobelias. *Evolutionary Ecology* 4:157–171.

Zhang, S.-Y. 1995. Activity and ranging patterns in relation to fruit utilization by brown capuchins (*Cebus apella*) in French Guiana. *International Journal of Primatology* 16:489–507.

Zhou, J., Li, W.-B., Wang, X., and Li, J.-H. 2022. Seasonal change in activity rhythms and time budgets of Tibetan macaques. *Biology* 11:1260. DOI:10.3390/biology11091260

Zhou, Q., Wei, F., Huang, C., Li, M., Ren, B., and Luo, B. 2007. Seasonal variation in the activity patterns and time budgets of *Trachypithecus franscoisi* in the Nonggang Nature Reserve, China. *International Journal of Primatology* 28:657–671.

Zhou, Q., Wei, H., Tang, H., Huang, Z., Krston, A., and Huang, C. 2014. Niche separation of sympatric macaques, *Macaca assamensis* and *M. mulatta*, in limestone habitats of Nonggang, China. *Primates* 55:125–137.

Zuberbühler, K. 2000. Interspecific semantic communication in two forest primates. *Proceedings of the Royal Society B: Biological Sciences* 267:713–718.

Zuberbühler, K. 2001. Predator-specific alarm calls in Campbell's monkeys, *Cercopithecus campbelli*. *Behavioral Ecology and Sociobiology* 50:414–422.

Zuberbühler, K., and Jenny, D. 2002. Leopard predation and primate evolution. *Journal of Human Evolution* 43:873–886.

Zuberbühler, K., Jenny, D., and Bshary, R. 1999. The predator deterrence function of primate alarm calls. *Ethology* 105:477–490.

INDEX

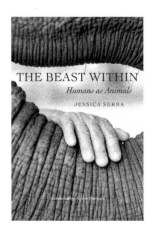

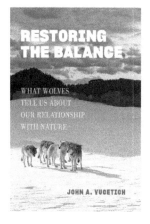
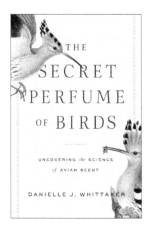

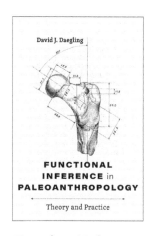

Milton Keynes UK
Ingram Content Group UK Ltd.
UKHW021900310824
447561UK00002B/5